Insecticides - Advances in Insect Control and Sustainable Pest Management

Edited by Habib Ali, Adnan Noor Shah, Muhammad Bilal Tahir, Sajid Fiaz and Basharat Ali

Published in London, United Kingdom

Insecticides - Advances in Insect Control and Sustainable Pest Management
http://dx.doi.org/10.5772/intechopen.104285
Edited by Habib Ali, Adnan Noor Shah, Muhammad Bilal Tahir, Sajid Fiaz and Basharat Ali

Contributors
François Essouma Manga, Mvondo Nganti Dorothée, Victorine Obe Lombeko, Katya Francine Erica Emvoutou, Zachée Ambang, Rabbiah Manzoor Malik, Sahar Fazal, Narjis Khatoon, Saima Batool, Muneeba Ishtiaq, Syed Tauqeer Abbas, Melis Yalçın, Banka Kanda Kishore Reddy, Addanki Maneesha, Malleswari Sadhineni, Chinna Babu Naik, Tejaswi Yelleti, G. Raja Reddy, Temesgen Fita Fita, Emana Getu, Mulatu Wakgari, Kebede Woldetsadik, Jesús Antonio Salazar-Magallón, Hugo Rodríguez-Romero, Laura Rodríguez-Pelaéz, Adriana Reyes-Castro, Orlando Tlakaele Notario-Rendón, Manuel González-Pérez, Juventino Reza-Salgado, Pedro Antonio Rodríguez-Salazar, Shikha Gaikwad, Bouzenna Hafsia, Najla Hfaiedh, Hadil Kadri, Dorsaf Ben Jeddou, Abassi Raoudha, Guesmi Fatma

First published in London, United Kingdom, 2023 by IntechOpen
IntechOpen is the global imprint of INTECHOPEN LIMITED, registered in England and Wales, registration number: 11086078, 5 Princes Gate Court, London, SW7 2QJ, United Kingdom

British Library Cataloguing-in-Publication Data
A catalogue record for this book is available from the British Library

Additional hard and PDF copies can be obtained from orders@intechopen.com

Insecticides - Advances in Insect Control and Sustainable Pest Management
Edited by Habib Ali, Adnan Noor Shah, Muhammad Bilal Tahir, Sajid Fiaz and Basharat Ali
p. cm.
Print ISBN 978-1-80356-419-7
Online ISBN 978-1-80356-420-3
eBook (PDF) ISBN 978-1-80356-421-0

For EU product safety concerns:
IN TECH d.o.o., Prolaz Marije Krucifikse Kozulić 3, 51000 Rijeka, Croatia, info@intechopen.com or visit our website at intechopen.com.

Meet the editors

Dr. Habib Ali, Ph.D., is an assistant professor at Khwaja Fareed University of Engineering and Information Technology, Pakistan. He graduated with a specialization in Agricultural Entomology from the University of Agriculture in Faisalabad, Pakistan, in 2012. He obtained a Ph.D. in Agricultural Entomology and Insect Pest Control from Fujian Agriculture and Forestry University (FAFU), China, in 2018. He is passionate about research work and has published various national and international research papers in high-impact journals. He has also edited several books and book chapters. Dr. Ali received the Best International Student of the Year award and the Best Thesis award from the China Scholarship Council (CSC). He was nominated for the 2021 World Scientist Index and awarded a Council of Asian Science editorship. He has attended many international conferences/symposiums/workshops, educational and faculty training, and completed national and international research projects. Moreover, Dr. Ali is a journal editor and reviewer. His areas of research interests include agricultural entomology, molecular entomology, bee health and disease management, crop pests, biological control of pests, microbial symbiosis, *Wolbachia*–infect interaction, biopesticides, biocontrol agents, and integrated pest management.

Dr. Sajid Fiaz is an assistant professor in the Department of Plant Breeding and Genetics, University of Haripur, Pakistan. Dr. Fiaz received his Ph.D. in Crop Genetics and Breeding from the Chinese Academy of Agricultural Sciences in 2019. Dr. Fiaz's research interests include marker-assisted selection for agronomic traits, mutation breeding for biotic and abiotic stress resistance, QTL mapping, and genome editing for yield and quality traits in cereals. Dr. Fiaz is actively supervising undergraduate and post-graduate students. He has been awarded with a Start-Up Research Grant Project from the Higher Education Commission of Pakistan worth 1 million PKR. He has published more than seventy-four research papers, six review articles, and ten book chapters. He is an academic editor for *PLOS One*, guest associate editor for *Frontiers in Plant Sciences*, review editor for *Frontiers in Genetics*, and is editing a special issue of *Plants*. Currently, Dr. Fiaz's lab group is working to identify genetic factors controlling drought stress tolerance in exotic and local landraces of rice.

Dr. Muhammad Bilal Tahir is an assistant professor, department head, the director of a high-tech laboratory, and the director of the Office of Research Innovation and Commercialization, Khwaja Fareed University of Engineering and Information Technology, Pakistan. His research interests include energy and environmental applications, material science and nanotechnology, environmental engineering, and photocatalysis. He is listed among the top 2% of scientists worldwide. He has published more than 800 research papers on enhanced oil recovery (EOR), renewable energy, alternative fuels, waste-to-energy, catalytic and resource recovery. Dr. Tahir is a guest editor of several special journal issues.

Dr. Adnan Noor Shah obtained a Ph.D. in Agronomy from Huazhong Agriculture University, Wuhan, China, in 2017. His specializations include plant nutrition, plant–soil interaction, crop physiology, crop cultivation and farming systems, crop ecology, plant abiotic stress responses, plant genomics, and epigenetics. From 2018 to 2019 he was an assistant professor at Gomal University Dera Ismail Khan, Pakistan. He completed a post-doctoral fellowship at Anhui Agricultural University, China in 2021. Currently, he is an assistant professor in the Department of Agricultural Engineering, Khwaja Fareed University of Engineering and Information Technology (KFUEIT), Pakistan. He is an editorial board member for different international and national journals, including *BMC Plant Biology*, *International Journal of Plant Production*, *Functional Plant Biology*, *Frontiers in Plant Science*, *Frontiers in Agronomy*, *Journal of Pure and Applied Agriculture*, and *Agriculture*.

Dr. Basharat Ali is an Assistant Professor of Agronomy at the Department of Agricultural Engineering, Khwaja Fareed University of Engineering and Information Technology (KFUEIT), Pakistan. He obtained a BSc (Hons) in Agriculture-Agronomy and an MSc (Hons) in Agronomy from the University of Agriculture Faisalabad, Pakistan, in 2009 and 2011, respectively. He obtained a Ph.D. in Agronomy from Zhejiang University, China, in 2014. His specializations include abiotic stress tolerance, plant nutrition, and plant growth regulation. After obtaining his Ph.D., Dr. Ali completed a post-doctoral fellowship at Zhejiang University, China, and an Alexander von Humboldt (AvH) fellowship from the University of Bonn, Germany. Dr. Ali was an assistant professor (IPFP and visiting faculty) in the Department of Agronomy, University of Agriculture Faisalabad from 2018 to 2022. He has published more than 100 journal articles and contributed to several international books.

Contents

Preface

In recent years, the field of insect control and pest management has experienced significant advancements and transformations. The increasing challenges posed by insect pests in various sectors, such as agriculture, public health, and urban environments, have necessitated the development of innovative and sustainable solutions. This book, *Insecticides – Advances in Insect Control and Sustainable Pest Management*, provides a comprehensive overview of the latest scientific progress and practical applications within this rapidly evolving field.

Insects, with their remarkable diversity and adaptability, play crucial roles in Earth's ecosystems. However, their ability to disrupt human activities and cause economic losses should not be underestimated. Traditional methods of insect control, often reliant on chemical insecticides, have often led to unintended consequences such as environmental pollution, harm to non-target organisms, and the emergence of insecticide resistance. Consequently, there is an urgent need to explore alternative approaches that effectively manage insect pests while minimizing adverse effects on the environment and human health.

This book brings together a distinguished group of experts, each contributing their unique perspectives and expertise, to address the multifaceted challenges of insect control. The chapters in this volume encompass a broad range of topics, including insect biology and behavior, innovative insecticide formulations, biopesticides, integrated pest management strategies, and the social, economic, and regulatory dimensions of insecticide usage. By examining current research and successful case studies, readers will gain insights into promising advancements and sustainable practices that are shaping the future of insect control.

The chapters are organized to provide a comprehensive understanding of insecticides, their modes of action, and their integration into holistic pest management approaches. We delve into the intricacies of insect physiology, genetics, and ecology to illuminate the underlying mechanisms that influence their behavior and adaptations. Furthermore, we explore the development and application of novel chemical compounds, biological agents, and cutting-edge technologies that have the potential to revolutionize insect control and reduce reliance on traditional insecticides.

We recognize that insect control is a dynamic field, continuously yielding new discoveries. Therefore, this book serves as a foundation for further exploration and innovation in insect control and sustainable pest management. Our hope is that the knowledge shared within these pages will inspire researchers, practitioners, and policymakers to collaborate and develop effective strategies that protect our crops, safeguard public health, and conserve our precious ecosystems.

We extend our sincere appreciation to all the contributors who have generously shared their expertise and insights to make this book possible. Their collective efforts have resulted in a valuable resource that addresses current challenges and provides a roadmap for a sustainable future in insect control.

We invite you to embark on a journey through the captivating world of insecticides and sustainable pest management. Together, let us strive to strike a balance between effective pest control and the preservation of our environment for generations to come.

Habib Ali, Adnan Noor Shah, Muhammad Bilal Tahir and Basharat Ali
Department of Agricultural Engineering,
Khwaja Fareed University of Engineering and Information Technology,
Rahim Yar Khan, Pakistan

Dr. Sajid Fiaz
Department of Plant Breeding and Genetics,
University of Haripur,
Haripur, Pakistan

Section 1

Pest Control: In Ecological Prospective

Chapter 1

Use of Biological and Chemical Pesticides in Agricultural Production: What Fate for Entomopathogenic Fungi?

François Essouma Manga, Mvondo Nganti Dorothée, Victorine Obe Lombeko, Katya Francine Erica Emvoutou and Zachée Ambang

Abstract

In the context of integrated pest management, the compatibility between the products used and even with the natural enemies of crop pests should still be elucidated. In this study, interviews were done with about 60 coffee growers to assess the use of pesticides in the protection of coffee berries. Then, *in vitro* tests were used to evaluate the effects of extracts of the seed powders of *Thevetia peruviana*, *Azadirachta indica*, the chlorpyriphos-ethyl insecticide and the chlorothalonil + dimethomorph fungicide, on the development parameters of the entomopathogenic fungus *Beauveria bassiana*, the natural enemy of the coffee berry borer, *Hypothenemus hampei*. The said tests consisted of the method of poisoning the culture medium with pesticides, observation and counting of spores under the optical microscope. The analysis of the collected data showed that depending on the type and severity of the pest pressure, growers apply several types of mainly chemical pesticides. Among the pesticides tested, extracts of *T. peruviana*, *A. indica* and chlorpyriphos-ethyl considerably reduced the development of *B. bassiana*. These results show that in crop protection, the use of biological or chemical substances should be done in a judicious way, to ensure the conservation and the valorization of natural enemies of phytosanitary pressures.

Keywords: compatibility, *Beauveria bassiana*, pesticide plants, chemical pesticides, integrated control

1. Introduction

Some studies point out that today's, food and agricultural systems have succeeded in supplying large quantities of food to global markets with high use of chemical inputs, but are degrading land, water, ecosystems, biodiversity and human health [1]. To address the increasing use of chemical pesticides in agricultural production, a fundamentally different model of agriculture is now required, one that reduces or rejects chemical inputs, optimizes biodiversity, and stimulates interactions between different species [2]. Pesticidal plants and entomopathogenic fungi, both gifts of nature, are effective alternatives to chemical pesticides in the integrated management of crop pests and diseases [3–6]. In this sense, several authors have already proven the effectiveness of plants of the genera *Azadirachta* (Neem) and *Thevetia* (Yellow Laurel) [7, 8], as well as biological control agents of the genera *Beauveria* and *Metarhizium* [9–12]. However, the introduction of these exotic pests is the most widely used approach in agriculture, at the expense of conserving and valorizing existing biological control agents in the agroecosystem [13–16], and without verification of their compatibility.

Given the effectiveness of pesticidal plant extracts and biological pest control agents in pest management, could they be used synergistically? This probably requires compatibility [17] which has been assessed so far by a few disparate studies [18–37]. These cited studies focus on the compatibility or synergy between chemical pesticides and entomopathogenic fungi. In view of the rareness of studies on the compatibility between biological pesticides (pesticidal plants and entomopathogenic fungi), and our previous research on the use of *B. bassiana* and extracts of *Thevetia peruviana* and *Azadirachta indica* in the control of the coffee berry borer (*Hypothenemus hampei* Fer.), this study was initiated to provide some information. Specifically, the aim is to evaluate the compatibility between extracts of *T. peruviana* and *A. indica* seed powders, the chlorpyrifos-ethyl insecticide and the chlorothalonil + dimethomorph fungicide, on the development parameters of the entomopathogenic fungus *B. bassiana*, in the control of phytosanitary pressures on coffee.

2. Methodology

The biological material consisted of *Thevetia peruviana* seed powder, *Azadirachta indica* seed powder and oil, and two isolates of *Beauveria bassiana*. Both strains were isolated from composite soil samples of Arabica (Bb-IRAD.Fbt) and Robusta (Bb-IRAD.Nkoe) coffee plantations [12], and stored at the Central Laboratory of Phytopathology (CLP) of the Mbalmayo Agricultural Research Centre (MARC) of the Institute of Agricultural Research for Development (IRAD), Cameroon. Compatibility tests were carried out in the same laboratory. The effects of the plant extracts on *B. bassiana* were compared to those caused by synthetic pesticides. These synthetic pesticides consisted of the chemical insecticide, Pyriforce composed of chlorpyriphos-ethyl 600 g/l as active substance, and the chemical fungicide, Sphinx composed of chlorothalonil 400 g/kg + dimethomorph 80 kg as active substances.

The study was carried out in the localities of Melong, Bamendjou and Doumé where the producers surveyed were selected. These localities were chosen because they are among the major Arabica (Bamendjou) and Robusta (Melong and Doumé) coffee production basins in Cameroon, are accessible by national roads and are located in three different agroecological zones. At the farm level, farmers were surveyed to assess pest pressures and control strategies.

In the CLP and the Central Laboratory of Entomology (CLE) of the IRAD, insect rearing, fungus isolation and efficacy tests on the bark beetle and compatibility tests between plant extracts and *B. bassiana* were performed.

2.1 Evaluation of pest and disease control strategies in coffee farms

The different strategies for regulating phytosanitary constraints and improving yields were evaluated through surveys. The experimental unit consisted of sixty-three farmers surveyed, divided into twenty-two farmers in Bamendjou, twenty-one farmers in Melong and twenty farmers in Doumé. The selection of farmers was done using the referral sampling technique (non-probability sampling method), where a farmer refers to an individual with a farm. Thus, farmers with at least one dependent coffee farm were chosen as the basic sampling unit. Semi-structured comprehensive interviews were conducted using a pre-established open-ended questionnaire. The questionnaire included semi-structured questions that allowed for the collection of information on diseases and pests affecting coffee farms, as well as strategies and techniques to control these pests.

2.2 Obtaining the different concentrations of biological and chemical pesticides

The plant extracts were applied to each strain on the basis of four concentrations: C_1 (12.5), C_2 (25), C_3 (50) and C_4 (100) in mg/ml for the aqueous extracts and in µl/ml for the oil. The plant extracts were prepared and the tested doses were obtained according to the method used in previous studies [8, 38, 39]. The choice of used doses was based on the proof of their efficacy tested in some studies on *Phytophtora megakarya, Sahbergella singularis*, *Lasiodiplodia theobromae* and *Fusarium* sp. [7, 40]. Each treatment in the trial was repeated five times. The chemical fungicide and insecticide were used as positive controls (C_{0+}) at the manufacturers' recommended doses. The negative/absolute control (C_{0-}) was simply the PDA culture medium.

The recommended doses for the chemical insecticide and fungicide are 1 l/ha (50 ml for a 16-liter sprayer) and 3.33 g/l water, respectively. Petri dishes were prepared by taking 7 ml of the stock solution of each pesticide, and mixing it with 93 ml of PDA medium for a final volume of 100 ml of each product. This final volume was poured into five Petri dishes serving as replicates, 20 ml per dish. The different Petri dishes were placed in an incubation room under conditioned air at a temperature of T = 25°C and a humidity of ψ = 60%.

2.3 Effect of treatments on germination of *Beauveria bassiana*

To assess spore germination, 10 ml of spore solution was prepared by mixing spores from pure cultures of *B. bassiana* (twenty-one days old) with sterile distilled water and 1% tween 80. After homogenization with a magnetic stirrer, the solution was calibrated using the Malassez cell at the concentration of 1×10^6 spores/ml [41]. Then, three drops of each solution were individually placed in three different locations of five Petri dishes containing PDA medium, and covered with a coverslip. The plant extracts were calibrated at four concentrations (C_1, C_2, C_3 and C_4), the positive controls at one concentration each (C_{0+1}, C_{0+2}) and the absolute/negative control (C_{0-}). For each strain (Bb-IRAD.Nkoe and Bb-IRAD.Fbt), the five prepared Petri dishes were incubated in the dark at room temperature for 16 hours. After incubation, the different plates were observed under a light microscope for the enumeration of germinated and ungerminated spores; this was repeated three times. Any spore with an elongated germ tube was considered as germinated and viable spore. The germination rate of the spores was calculated using the following formula from [42]:

$$GR\ (\%) = \frac{A}{A+B} \times 100 \quad (1)$$

where: *GR* = spore germination rate; *A* = number of germinated spores; *B* = number of ungerminated spores. d1 + d22.

2.4 Effect of treatments on radial growth of *Beauveria bassiana*

To assess the radial growth of *B. bassiana* isolates, a 6 mm diameter mycelial disc was taken (from 21-day-old pure cultures of Bb-IRAD.Nkoe and Bb-IRAD.Fbt) and placed in the center of each Petri dish containing media supplemented with the different plant extracts and chemical pesticides. A negative control not supplemented with extracts and chemical pesticides was also prepared. Each treatment was repeated 5 times, each repetition corresponding to one Petri dish. Petri dishes were incubated at a temperature of T = 25°C, a humidity of ψ = 60% under a photoperiod of 12/12 and for 21 days. Using the perpendicular line method, each diameter or line was measured daily. The average of the two perpendicular measurements, subtracted from the diameter of the starting explant, was the measure of radial growth of the fungus. It was obtained using the following formula [43]:

$$RG = \frac{d1+d2}{2} - d0 \quad (2)$$

where: *RG* = Radial Growth; d_1 = first growth diameter (cm); d_2 = second growth diameter (cm); d_0 = diameter of the deposited explant (cm).

2.5 Effect of treatments on *Beauveria bassiana* spore production

The Petri dishes used in the evaluation of radial growth of *B. bassiana*, were used to evaluate the effects of treatments on spore production. Thus, a quantity of 10 ml of spore suspension of each isolate and each concentration of the different treatments was prepared as in the evaluation of the germination of isolates. In order to quantify the number of conidia produced by the fungus, five 1 ml samples of the suspension were successively taken from each dish and placed on the Malassez cell. The conidia were counted under a light microscope and the average number of conidia per observation was recorded.

2.6 Correlative and comparative assessment of compatibility between biological pesticides, chemical pesticides and *Beauveria bassiana*

The correlative and comparative assessment of compatibility was done using the percentages of inhibition or reduction of mycelial growth, germination and spore production of *B. bassiana* [44]. These percentages were calculated according to the following formula [45]:

$$IP\ or\ RR\ (\%) = \frac{Vt - Vx}{Vt} \times 100 \tag{3}$$

where: *IP* or *RR* = inhibition percentage or reduction rate; *Vt* = value of growth diameter, germination rate or number of conidia produced estimated on control medium; *Vx* = value of growth diameter, germination rate or number of conidia produced estimated in the presence of the extract or fungicide tested.

2.7 Statistical analysis

The experimental set-up adopted for the incubation of the Petri dishes was a completely randomized set-up with: two strains of *B. bassiana*; two aqueous extracts (aqueous extract of *A. indica* and *T. peruviana* seed powder) and an oil cold extract of *A. indica* seeds; and two chemical pesticides.

Microsoft Excel spreadsheet software was used to create the databases and XLSTAT 2014 software was used for statistical analysis. The experimental test data were first subjected to a Shapiro–Wilk normality test [46], followed by a square root transformation (radial growth and number of spores germinated or produced) or an ArcSin angular transformation (inhibition rate) [47, 48]. The sampled and transformed data underwent descriptive analysis, a general linear regression model with analysis of variance (ANOVA), followed by the multiple comparison test of means at 5% risk (α).

After these calculations, the inhibitory effect of the tested products was classified firstly by means of hierarchical ascending classification (HAC). This allowed treatments with the same effect to be grouped together. Secondly, the different groups were categorized on the basis of a list of inhibition levels [49], where:

- very toxic = >80% inhibition;
- toxic = 10–79% inhibition;
- low toxic or compatible = <10% inhibition.

At the end of each analysis, all interpretations and conclusions were drawn at the transformed scale, but the results presented were converted back to the original units [48, 50].

3. Results

3.1 Weed, disease and pest control strategies

Clearing of coffee fields and structural/sanitary pruning of coffee trees were carried out by 100% of farmers in each locality. Insecticides were applied by 100%, 80% and 59.1% of farmers in Melong, Doumé and Bamendjou, respectively. Fungicides were applied in Bamendjou by a significantly higher proportion of farmers (86.36%) than in other localities such as Doumé (5% of farmers). Fertilizers were widely applied in Melong and Bamendjou by 85.71 and 54.55% of farmers, respectively. Herbicides were applied in all localities by more than half of the farmers, but this percentage was significantly lower (59.1% of farmers) in Bamendjou (Fisher's exact test at 0.05 significance level) (**Figure 1**).

3.2 Effects of treatments on *Beauveria bassiana* spore germination

The results showed that spore germination of both *B. bassiana* isolates was reduced by the different treatments and concentrations used. Compared to the absolute control, the percentage of germination was higher in Bb-IRAD.Nkoe than in Bb-IRAD.Fbt at concentrations C_1, C_2 and C_3, respectively for reduction rates ranging from 54 to 86% for AEAI on Bb-IRAD.Nkoe against 46 to 84% for AETP on Bb-IRAD.Fbt. Thus, at C_4 concentrations of these extracts, reduction rates of 100% were observed. This rate is identical to that obtained with chlorpyrifos-ethyl, hence no significant difference (**Figure 2B**).

With the *A. indica* oil extract (OEAI), the germination rate was lower for Bb-IRAD.Nkoe than for Bb-IRAD.Fbt at concentrations C_1 and C_2 (86 and 77% versus 89 and 85% respectively). Furthermore, this extract totally inhibited germination (0% germination rate) of spores at concentrations C_3 and C_4. Nevertheless, Fisher's test showed no significant difference between germination rates at these concentrations and that of chlorpyrifos-ethyl (**Figure 2A**).

Tests with the chlorothalonil + dimethomorph fungicide showed little or no reduction in spore germination of both isolates. The lowest germination inhibition rates (7 and 6%) were recorded for Bb-IRAD.Nkoe and Bb-IRAD.Fbt, respectively. Nevertheless, Fisher's test showed significant differences between the germination inhibition rates recorded with this fungicide and those with the negative control (60.33 and 56.33% for Bb-IRAD.Nkoe and Bb-IRAD.Fbt, respectively) (**Figure 2A**).

Finally, with chlorpyrifos-ethyl, $AETP_4$, $AEAI_4$ and $OEAI_3$, a germination inhibition rate of 100% was recorded in both *B. bassiana* isolates. However, for the same concentrations, AETP had more effect on spore germination of both isolates than AEAI (**Figure 2B**).

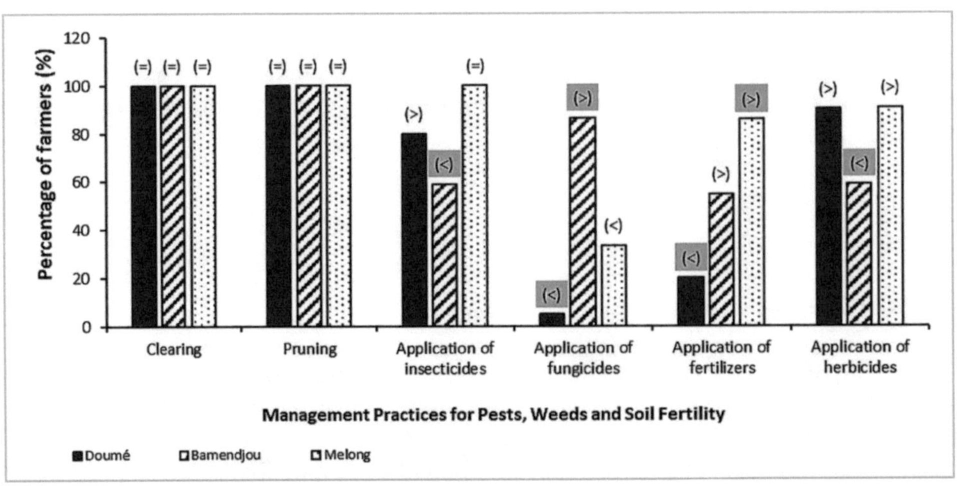

Figure 1.
Practices applied to manage pests and diseases in coffee farms. Values shown with the symbol < or > in bold are significant according to Fisher's exact test at the 0.05 significance level.

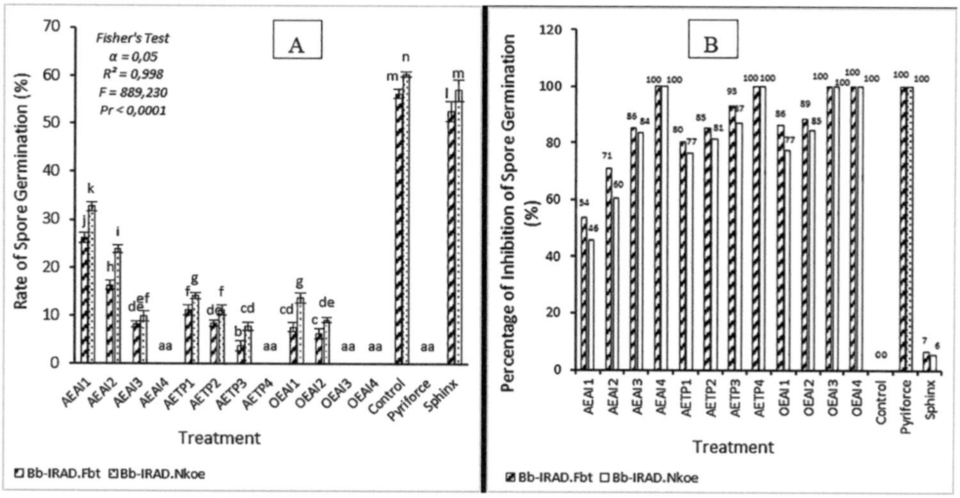

Figure 2.
Germination and germination inhibition rates of conidia under the influence of plant extract, chemical fungicide and insecticide treatments. AEAI = Azadirachta indica *aqueous extract; AETP =* Thevetia peruviana *aqueous extract; OEAI =* Azadirachta indica *oil extract; pyriforce = chemical insecticide composed of chlorpyrifos-ethyl; sphinx = chemical fungicide composed of chlorothalonil + dimethomorph; 1, 2, 3 and 4 correspond to C_1 (12.5), C_2 (25), C_3 (50) and C_4 (100) in mg/ml for the aqueous extracts and in µl/ml for the oil extract. *Means with the same letter are not significantly different at the 0.05 level according to Fisher's test.*

3.3 Effects of treatments on radial growth of *Beauveria bassiana* isolates

The aqueous extract of *A. indica* seed powder inhibited the growth of both isolates at all concentrations. This inhibition of radial growth was proportional and significant at the concentrations tested in both isolates (**Figure 3**). Thus, a total inhibition (100%) of radial growth was observed with $AEAI_4$ in both isolates and only in Bb-IRAD.Fbt with $AEAI_3$ (**Table 1**).

The inhibition of mycelial growth of *B. bassiana* isolates in the presence of AEAI and AETP was significant and proportional to the concentrations tested compared to the negative control (0 ± 0). This inhibition was relatively stronger in Bb-IRAD.Fbt with both types of aqueous extracts than in Bb-IRAD.Nkoe with the aqueous extract of *A. indica*. In contrast to the negative control, Fisher's test showed no significant difference between the two aqueous extracts and chlorpyrifos-ethyl at the $AETP_4$ concentration (**Table 1**).

With OEAI the inhibition of mycelial growth of isolates was more pronounced at all concentrations. No significant difference was observed between the oil extract of *A. indica* and chlorpyrifos-ethyl in the two *B. bassiana* isolates (**Table 1**).

With the Chlorothalonil + Dimethomorph complex, a very small reduction in the growth of both *B. bassiana* isolates was observed (**Figure 4**). Thus, low percentages of inhibition were observed (16.27 ± 3.50 and 21.03 ± 1.32%) in Bb-IRAD.Fbt and Bb-IRAD.Nkoe, respectively. However, Fisher's test showed a significant difference between these percentages of inhibition and that caused by the negative control (**Table 1**).

3.4 Effects of treatments on *Beauveria bassiana* spore production

The test results revealed that there was very little influence of Sphinx on spore production of both *B. bassiana* isolates. Consequently, sporulation reduction rates

Applied Treatment	Average growth diameter (cm)		Inhibition rate (%)	
	Beauveria bassiana isolates			
	Foumbot	**Nkoémvone**	**Foumbot**	**Nkoémvone**
$AEAI_1$	0.47 ± 0.25^{h}	2.14 ± 0.37^{k}	94.36 ± 2.94^{fg}	47.09 ± 1.49^{d}
$AEAI_2$	0.33 ± 0.20^{ef}	1.52 ± 0.33^{j}	96.48 ± 2197^{ghij}	64.10 ± 3.67^{e}
$AEAI_3$	0 ± 0^{a}	0.39 ± 0.19^{fg}	100 ± 0^{m}	94.18 ± 1.86^{fg}
$AEAI_4$	0 ± 0^{a}	0 ± 0^{a}	100 ± 0^{m}	100 ± 0^{m}
$AETP_1$	0.32 ± 0.17^{ef}	0.57 ± 0.26^{i}	96.31 ± 2.01^{ghij}	92.79 ± 3.35^{f}
$AETP_2$	0.20 ± 0.13^{cd}	0.36 ± 0.21^{fg}	98.08 ± 1.48^{ijkl}	96.39 ± 2.81^{ghi}
$AETP_3$	0.14 ± 0.1^{bc}	0.21 ± 0.13^{cd}	98.86 ± 1.08^{klm}	97.98 ± 1.57^{ijkl}
$AETP_4$	0 ± 0^{a}	0.08 ± 0.09^{ab}	100 ± 0^{m}	99.66 ± 0.86^{lm}
$OEAI_1$	0.12 ± 0.14^{bc}	0.41 ± 0.28^{gh}	99.46 ± 1.41^{klm}	96.20 ± 3.69^{gh}
$OEAI_2$	0.08 ± 0.09^{ab}	0.27 ± 0.19^{de}	99.66 ± 0.88^{lm}	97.73 ± 2.37^{hijk}
$OEAI_3$	0.05 ± 0.07^{ab}	0.18 ± 0.12^{c}	99.82 ± 0.59^{lm}	98.58 ± 1.60^{jkl}
$OEAI_4$	0 ± 0^{a}	0 ± 0^{a}	100 ± 0^{m}	100 ± 0^{m}
Control	4.22 ± 0.59^{n}	4.21 ± 0.60^{n}	0 ± 0^{a}	0 ± 0^{a}
Pyriforce	0 ± 0^{a}	0 ± 0^{a}	100 ± 0^{m}	100 ± 0^{m}
Sphinx	3.40 ± 0.53^{m}	3.31 ± 0.52^{l}	16.27 ± 3.50^{b}	21.03 ± 1.32^{c}
Analysis of Variance	$R^2 = 0.993$ F = 295.936 Pr < 0.0001		$R^2 = 0.985$ F = 133.447 Pr < 0.0001	

AEAI = Aqueous extract of Azadirachta indica*; AETP = Aqueous extract of* Thevetia peruviana*; OEAI = Oil extract of* Azadirachta indica*; Pyriforce = Insecticide composed of Chlorpyriphos-ethyl; Sphinx = Fungicide composed of Chlorothalonil + Dimethomorph; 1, 2, 3 and 4 correspond to C_1 (12.5), C_2 (25), C_3 (50) and C_4 (100) in mg/ml for the aqueous extracts and in µl/ml for the oil extract. * Means with the same letter are not significantly different at the 0.05 level according to Fisher's test.*

Table 1.
Radial growth and inhibition rate of Beauveria bassiana isolates according to treatments after 21 days of incubation.

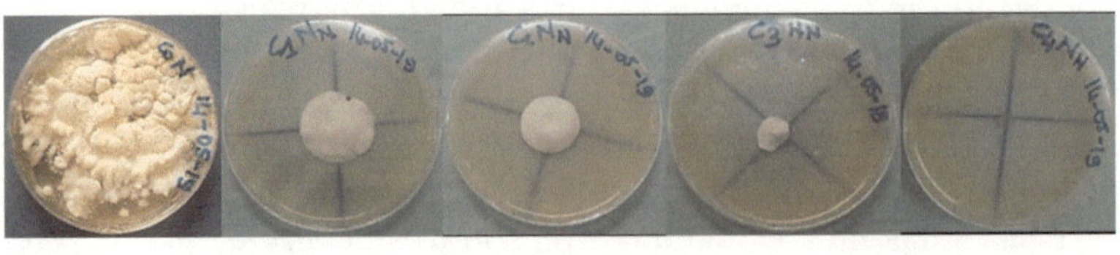

Figure 3.
Inhibition of mycelial growth of bb-IRAD.Nkoe by aqueous extract of Azadirachta indica *on day 21 of growth. N = neem or* Azadirachta indica*; $_N$ = Nkoemvone isolate; 1, 2, 3 and 4 correspond to C_1 (12.5), C_2 (25), C_3 (50) and C_4 (100) in mg/ml for the aqueous extracts and in µl/ml for the oil extract; C_o = sterile distilled water + tween 80.*

were higher with Bb-IRAD.Fbt than with Bb-IRAD.Nkoe regardless of the treatment. The aqueous extracts (AEAI and AETP) of both plants caused 100% reduction rates in both isolates only with C_4, while 100% reduction was observed with $OEAI_3$ and $OEAI_4$. However, all treatments except the negative control caused more than 50% reduction in sporulation in both isolates (**Figure 5B**).

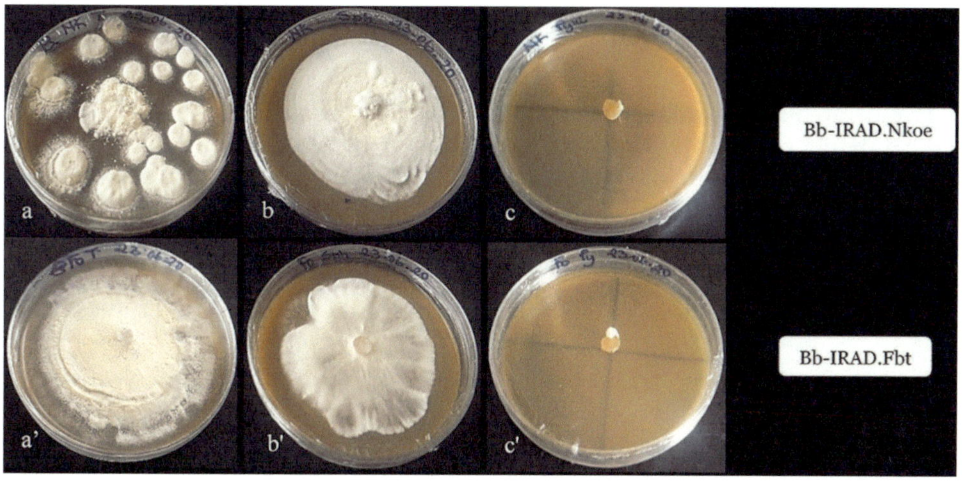

Figure 4.
Mycelial growth of the two isolates of Beauveria bassiana under the effect of sphinx and pyriforce. A and a' = control; b and b' = sphinx (chlorothalonil + dimethomorph); c and c' = pyriforce (chlorpyriphos-ethyl).

Finally, spore production in both *B. bassiana* isolates was relatively low with all OEAI concentrations. The absence of spore production was noted with $OEAI_3$ and $OEAI_4$ in both isolates. Bb-IRAD.Fbt showed spore production with $OEAI_1$ and $OEAI_2$, but this was not significant compared to that with $OEAI_3$, $OEAI_4$ and pyriforce (**Figure 5A**).

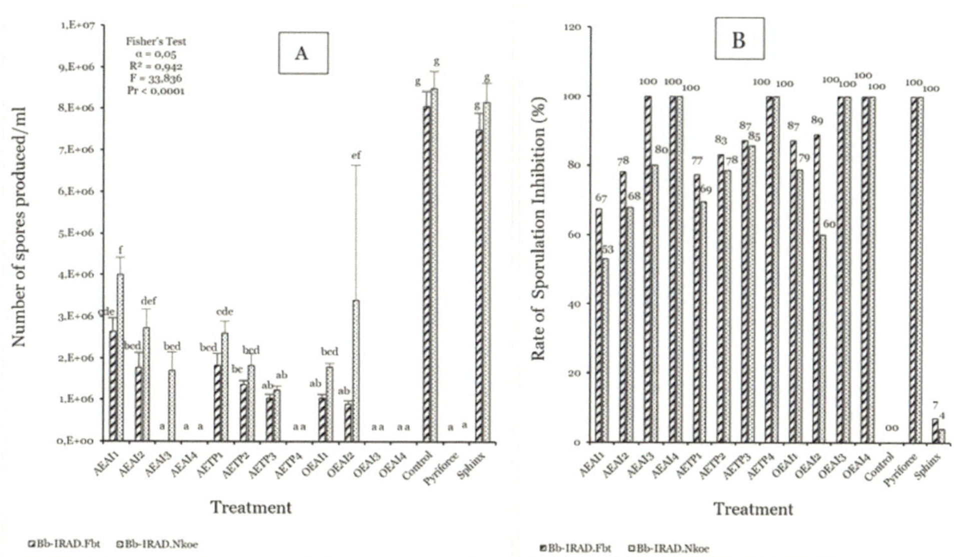

Figure 5.
Average number of conidia produced and sporulation inhibition rate in each treatment. AEAI = aqueous extract of Azadirachta indica; *AETP = aqueous extract of* Thevetia peruviana; *OEAI = oil extract of* Azadirachta indica; *pyriforce = insecticide composed of chlorpyriphos-ethyl; sphinx = fungicide composed of chlorothalonil + dimethomorph; 1, 2, 3 and 4 correspond to C_1 (12.5), C_2 (25), C_3 (50) and C_4 (100) in mg/ml for the aqueous extracts and in µl/ml for the oil extract. *means with the same letter are not significantly different at the 0.05 level according to Fisher's test.*

3.5 Relationship between treatments and inhibition of *Beauveria bassiana* development

The hierarchical ascending classification (HAC) of the inhibition rates of *B. bassiana* by the different treatments shows that all the treatments, AETP, OEAI, pyriforce and AEAI were very toxic (inhibition between 98.31 or 92.79%) for the radial growth of both isolates. Only $AEAI_1$ and $AEAI_2$ were just toxic (inhibition between 64.10 and 47.09%) to Bb-IRAD.Nkoe. Water and sphinx were compatible with both isolates (**Table 2**).

Classification variable	Classes	Barycentre (%) *[Interval]*	Treatment-Isolate	Classification status
Inhibition rate of radial growth (%)	1	98,190 [0,117—5398]	$AEAI_1$-Fbt $AEAI_2$-Fbt $AEAI_3$-Fbt $AEAI_4$-Fbt $AETP_1$-Fbt $AETP_2$-Fbt $AETP_3$-Fbt $AETP_4$-Fbt $OEAI_1$-Fbt $OEAI_2$-Fbt $OEAI_3$-Fbt $OEAI_4$-Fbt Pyriforce-Fbt $AEAI_3$-Nk $AEAI_4$-Nk $AETP_1$-Nk $AETP_2$-Nk $AETP_3$-Nk $AETP_4$-Nk $OEAI_1$-Nk $OEAI_2$-Nk $OEAI_3$-Nk $OEAI_4$-Nk Pyriforce-Nk	Very Toxic
	2	9325 [6942—11,709]	Control-Fbt Sphinx-Fbt Control-Nk Sphinx-Nk	Compatible
	3	55,597 [8507—8507]	$AEAI_1$-Nk $AEAI_2$-Nk	Toxic
Inhibition rate of spore germination (%)	1	53,301 [0,249—7445]	$AEAI_1$-Fbt $AEAI_1$-Nk $AEAI_2$-Nk	Toxic
	2	90,484 [1726—19,478]	$AEAI_2$-Fbt $AEAI_3$-Fbt $AEAI_4$-Fbt $AETP_1$-Fbt $AETP_2$-Fbt $AETP_3$-Fbt $AETP_4$-Fbt $OEAI_1$-Fbt $OEAI_2$-Fbt $OEAI_3$-Fbt $OEAI_4$-Fbt	Very Toxic

Classification variable	Classes	Barycentre (%) *[Interval]*	Treatment-Isolate	Classification status
			Pyriforce-Fbt $AEAI_3$-Nk $AEAI_4$-Nk $AETP_1$-Nk $AETP_2$-Nk $AETP_3$-Nk $AETP_4$-Nk $OEAI_1$-Nk $OEAI_2$-Nk $OEAI_3$-Nk $OEAI_4$-Nk Pyriforce-Nk	
	3	3008 [2516—3500]	Control-Fbt Sphinx-Fbt Control-Nk Sphinx-Nk	Compatible
Inhibition rate of Sporulation (%)	1	76,130 [1143—23,189]	$AEAI_1$-Fbt $AEAI_2$-Fbt $AETP_1$-Fbt $AETP_2$-Fbt $AETP_3$-Fbt $OEAI_1$-Fbt $OEAI_2$-Fbt $AEAI_1$-Nk $AEAI_2$-Nk $AEAI_3$-Nk $AETP_1$-Nk $AETP_2$-Nk $AETP_3$-Nk $OEAI_1$-Nk $OEAI_2$-Nk	Toxic
	2	100,000 [0–0]	$AEAI_3$-Fbt $AEAI_4$-Fbt $AETP_4$-Fbt $OEAI_3$-Fbt $OEAI_4$-Fbt Pyriforce-Fbt $AEAI_4$-Nk $AETP_4$-Nk $OEAI_3$-Nk $OEAI_4$-Nk Pyriforce-Nk	Very Toxic
	3	2737 [1185—4288]	Control-Fbt Sphinx-Fbt Control-Nk Sphinx-Nk	Compatible

Fbt = Bb-IRAD.Fbt; Nk = Bb-IRAD.Nkoe; AEAI = Aqueous extract of Azadirachta indica*; AETP = Aqueous extract of* Thevetia peruviana*; OEAI = Oil extract of* Azadirachta indica*; Pyriforce = Insecticide composed of Chlorpyriphos-ethyl; Sphinx = Fungicide composed of Chlorothalonil + Dimethomorph; 1, 2, 3 and 4 correspond to C_1 (12.5), C_2 (25), C_3 (50) and C_4 (100) in mg/ml for the aqueous extracts and in µl/ml for the oil extract.*

Table 2.
Hierarchical ascending classification of treatments according to their inhibition of growth parameters of Beauveria bassiana isolates.

Spores germination of the isolates was impacted by most of the treatments AETP, OEAI, pyriforce and AEAI which were highly toxic (90% inhibition on average) except $AEAI_1$ for Bb-IRAD.Fbt, $AEAI_1$ and $AEAI_2$ for Bb-IRAD.Nkoe which were toxic (50% inhibition on average) for these isolates. Water and sphinx were compatible with spore germination (**Table 2**).

As for sporulation, the high concentrations 3 and 4 of OEAI (for both isolates), 3 and 4 of AEAI (for Bb-IRAD.Fbt), 4 of AETP (for both isolates) and pyriforce were very toxic by totally (100%) inhibiting spore production. The low concentration C_1 of all treatments was found to be toxic for both isolates, however, concentrations 3 of AETP (for both isolates) and AEAI (for Bb-IRAD.Fbt) were also in this toxic class with inhibition ranging from 77.27 to 52.94%. Sterile distilled water and sphinx were found to be compatible with both isolates (**Table 2**).

4. Discussion

Coffee farmers use several strategies to cope with pest pressures. These practices are applied or not depending on the constraints encountered, their period of occurrence, incidence and severity [51]. Unfortunately, this study found that all the products used to control pests were synthetic products. In all study sites, there is a need to use organic products for quality coffee production and to ensure protection of human and environmental health. However, the unavailability of organic products in the market, low knowledge of their use and historical dependence on synthetic products complicate the adoption of organic products [52] and inexorably push producers towards synthetic products.

The products tested in vitro, with the exception of chlorothalonil + dimethomorph, significantly reduced germination, mycelial growth and spore production of the *B. bassiana* isolates used. Thus, the application of this fungicide to control fungal diseases such as anthracnose of coffee berries, allows the conservation of the natural inoculum of *B. bassiana* in the field and a synergistic control of phytosanitary pressures on berries. This finding corroborates with the results of some works [53–55] who respectively showed that sulfur, copper oxychloride and strobilurin fungicides are compatible with *B. bassiana* isolates although their primary faculty is antifungal.

All OEAI, AETP and AEAI treatments were found to be toxic to both isolates. This toxicity of the extracts of both plants to *B. bassiana*, is thought to be due to terpenes, phenols, alcohols, alkaloids, tannins and other secondary metabolites (capable of inducing toxicity of cell walls, membranes and organelles [56, 57]. A conservation of *B. bassiana* spores remains hypothetical in the presence of these pesticidal plant extracts (PPE) because they prevent conidial germination, a very important step in pest control with CEP. Indeed, the onset of the epizootic is conditioned by the ability of these conidia to germinate on the host [58]. Similarly, the success of CEP depends on the viability of its spores [44], which is therefore threatened by PPE in this study. These results are similar to other studies [22, 59–62] which showed that *A. indica* oil extracts and azadirachtin 5EC (commercial biopesticide) at the recommended dose were incompatible with *B. bassiana*. Similarly, Margoside® (commercial formulation based on 0.3% neem oil) and neem extracts have been shown to delay in vitro spore germination of 23 isolates (out of 30 in total) of *B. bassiana*, but without significantly reducing it [23, 33].

However, results of some works have shown that neem oil (2.5%) and neem seed extracts, neem gold, Topneem, biospark and exodon (commercialized biological pesticides), show compatibility with all *B. bassiana* isolates obtained in these works [26, 32, 53, 63, 64]. Furthermore, some studies [65, 66] have shown that the combination of Azadirachtin (neem extract) with *B. bassiana*, has an additive effect. Similarly, a synergistic efficacy of *A. indica* leaf extracts and Azadirachtin (AzaMax; 200 ml × 100 l^{-1}) with *B. bassiana* has been proven respectively in the control of wheat aphids [67] and in the control of *Plutella xylostella* (L.) (Lepidoptera: Plutellidae) [68].

These contradictory results could be due to the qualitative and quantitative variability of the extracts used [69] by the different authors and to the genetic variability of the *B. bassiana* isolates used in this study [23]. It is therefore evident that the compatibility of plant protection products depends mainly on the nature of the compounds, the concentrations used and the nature of the isolates. Furthermore, studies have shown that commercial strains of *B. bassiana* are less resistant than wild strains possibly due to the effect of the products used for encapsulation/formulation [70, 71]. Further research on the effects of neem on the enzymatic activity of *B. bassiana* could be interesting to decide on these contradictions [72].

Apart from plant extracts, chlorpyrifos-ethyl at the recommended dose was also shown to be incompatible with *B. bassiana* by inhibiting all developmental stages of both isolates. This corroborates with some studies [53, 73–75] who reported that triazophos, chlorpyriphos and endosulfan formulations inhibited 100% of *B. bassiana* germination at all doses tested. This inhibition is due to the ability of the chemical insecticide to act as an acetylcholine esterase inhibiting neurotoxin [4].

Finally, a comparative look at the two isolates showed that Bb-IRAD.Fbt was more affected by EPP than Bb-IRAD.Nkoe. This further confirms that these two isolates are different hence their reactivity was variable to the applied pesticides. The different membrane and intracellular receptors of Bb-IRAD.Nkoe would be less specific to the toxic molecules AETP, AEAI and OEAI which act either at the membrane level (as a contact fungicide) or inside the cells (as a systemic fungicide) [76]. Therefore, these extracts can either inhibit metabolism by having a fungistatic effect or inhibit respiration by having a fungicidal effect on *B. bassiana* [7] resulting in incompatibility.

5. Conclusion

Cross-checking the results showed that all producers apply clearing and pruning, and disparately others apply insecticides, fungicides and herbicides. All the products used are chemical, with a large number for insecticides. AETP, OEAI, chlorpyrifos-ethyl and AEAI were found to be toxic to all developmental traits of *B. bassiana* isolates, with more than 50% inhibition at low and medium concentrations, and highly toxic at high concentrations, with 90% inhibition on average. However, the synthetic fungicide based on chlorothalonil + dimethomorph was found to be compatible with *B. bassiana* isolates as was the absolute control consisting of sterile distilled water.

This study shows that, although biological and with effects on *Hypothenemus hampei* or *Colletotrichum kahawae*, natural substances such as extracts of *T. peruviana* and *A. indica*, as well as chlorpyrifos-based products, do not allow either the preservation or the synergistic use of *B. bassiana* with extracts of these plants or chlorpyrifos-ethyl, especially at high concentrations of these substances. Therefore, these substances should be used with caution to ensure the sustainability and conservation of the diversity of natural enemies of coffee and other crop pests such as *B. bassiana*.

Acknowledgements

The authors are grateful to the International Foundation for Science (IFS) in Sweden [Research Grant Agreement No. I-1-C-6256-1] and the International Centre for Agricultural Research for Development (CIRAD) through the DP/Agroforestry in Cameroon, for their financial support. The authors are grateful to the Institute of Agricultural Research for Development (IRAD) for technical assistance.

Conflicts of interest

The authors have no conflict of interest to declare.

Author details

François Essouma Manga[1*], Mvondo Nganti Dorothée[2], Victorine Obe Lombeko[1], Katya Francine Erica Emvoutou[1] and Zachée Ambang[2]

1 Institut de Recherche Agricole pour le Développement (IRAD), Cameroon

2 University of Yaoundé, Department of Plant Biology, Cameroon

*Address all correspondence to: essouma2003@yahoo.fr

References

[1] Patil BV, Naik MK, Manujnath H, Hosamani AC. Chapter twenty-three: Formulation and compatibility of PGPR with pesticides for suppression of insect pests. In: Reddy MS, Ilao RO, Faylon PS, Dar WD, Batchelor WD, editors. Recent Advances in Biofertilizers and Biofungicides (PGPR) for Sustainable Agriculture. Cambridge Scholars Publishing; 2014. pp. 269-280

[2] Seema W. An overview on the biotechnological approaches in the control of crop pests. Journal of Eco-friendly Agriculture. 2006;**1**(2): 89-99

[3] Lacey LA, Frutos R, Kaya HK, Vail P. Insect pathogens as biological control agents: Do they have a future? Biological Control. 2001;**21**:230-248

[4] Lengai Geraldin MW, Muthomi James W, Mbega Ernest R. Phytochemical activity and role of botanical pesticides in pest management for sustainable agricultural crop production. Scientific African. 2020;**7** (2020):e00239. https://doi.org/10.1016/j.sciaf.2019.e00239

[5] Raguraman S, Kannan M. Non-target effects of botanicals on beneficial arthropods with special reference to *Azadirachta indica*. In: Singh D, editor. Advances in Plant Biopesticides. India: Springer; 2014. pp. 173-205. DOI: 10.1007/978-81-322-2006-0_10

[6] Lacey LA, Grzywacz D, Shapiro-Ilan DI, Frutos R, Brownbridge M, Goettel MS. Insect pathogens as biological control agents: Back to the future. Journal of Invertebrate Pathology. 2015;**132**:1-41

[7] Mboussi SB, Ambang Z, Ndogho PA, Ngoh Dooh JP, Manga Essouma F. In vitro antifungal potential of aqueous seeds extracts of *Azadirachta indica* and *Thevetia peruviana* against *phytophthora megakarya* in Cameroon. Journal of Applied Life Sciences International. 2016;**4**(4):1-12. DOI: 10.9734/JALSI/2016/23710

[8] Manga EF, Voula VA, Kone NAN, Mvondo Nganti D, Ambang Z. Coffee berry borer, *Hypothenemus hampei* Ferr.: Cultural control and *In vitro* effects of kernel extracts of *Azadirachta indica* A. Juss. And *Thevetia peruviana* (pers.) Schum. European Scientific Journal. 2021a;**17**(34):12-29

[9] Hajek AE, Delalibera I. Fungal pathogens as classical biological control agents against arthropods. Biological Control. 2010;**55**:147-158

[10] Mahot HC, Membang G, Hanna R, Begoude BAD, Bagny BL, Bilong BCF. Laboratory assessment of virulence of Cameroonian isolates of *Beauveria bassiana* and *Metarhizium anisopliae* against mirid bugs *Sahlbergella singularis* Haglund (Hemiptera: Miridae). African Entomology. 2019; **27**(1):86-96

[11] Membang G, Ambang Z, Mahot HC, Kuate AF, Fiaboe KKM, Hanna R. *Cosmopolites sordidus* (Germar) susceptibility to indigenous Cameroonian *Beauveria bassiana* (Bals.) Vuill. And Metarhizium anisopliae (Metsch.) isolates. Journal of Applied Entomology. 2020;**00**:1-13. https://doi.org/10.1111/jen.12757

[12] Manga EF, Ndo EGD, Voula VA, Mvondo ND, Assena LH, Ambang Z. Isolation of *Beauveria bassiana* (Deuteromycotina: Hyphomycetes) from the soils of coffee fields and insecticide activity against *Hypothenemus*

hampei (coleoptera: Scolytidae). Journal of Biopesticides. 2021b;**14**(2):141-153

[13] Vega FE, Infante F, Johnson AJ. Chapter 11: The genus *Hypothenemus*, with emphasis on *H. hampei*, the coffee berry borer. In: Vega FE, Hofstetter RW, editors. Bark Beetles: Biology and Ecology of Native and Invasive Species. London, UK: Elsevier/Academic Press; 2015. pp. 427-494. DOI: 10.1016/B978-0-12-417156-5.00011-3

[14] Aristizábal LF, Bustillo AE, Arthurs SP. Integrated pest management of coffee berry borer: Strategies from Latin America that could be useful for coffee farmers in Hawaii. Insects. 2016; 7(1):11-14

[15] Morris JR et al. Ant-mediated (Hymenoptera: Formicidae) biological control of the coffee berry borer: Diversity, ecological complexity, and conservation biocontrol. Myrmecological News. 2018;7(26):1-17

[16] Escobar-Ramírez a S, Ingo G, Inge A, Teja T. Biological control of the coffee berry borer_ Main natural enemies, control success, and landscape influence. Biological Control. 2019:17. DOI: 10.1016/j.biocontrol.2019.05.011

[17] Aldeghairi MA, El-Meleigi MA, Abdel Baky Nagdy F, Ibrahim Gamal H. The Entomopathogenic Fungus *Cladosporium* Sp. as a Candidate Biocontrol Agent against the Sweet Potato Whitefly, *Bemisia tabaci*, in Saudi Arabia. In: Reddy MS, Ilao RO, Faylon PS, Dar WD, Batchelor WD, Sayyed R, et al. editors. Recent Advances in Biofertilizers and Biofungicides (PGPR) for Sustainable Agriculture. Cambridge Scholars Publishing; 2014. pp. 481-509

[18] Shashi S, Gupta RBJ. Compatibility of *Beauveria brongniartii* with pesticides and organic manures. Pesticide Research Journal. 1998;**10**(2): 251-253

[19] Tamai MA, Alves SB, Lopes RB, Faion M, Padulla LFL. Toxicity of pesticides against *Beauveria bassiana* (Bals.) Vuill. Arquivos do Instituto Biológico. 2002;**69**(3):89-96

[20] Kouassi M, Coderre D, Todorova IS. Compatibility of zineb, dimethoate and *Beauveria bassiana* (Balsamo) Vuillemin against tarnished plant bug (Hemiptera: Miridae). Journal of Entomological Science. 2003;**38**(3):359-367

[21] Marques RP, Monteiro AC, Pereira GT. Growth, sporulation and viability of entomopathogenic fungi under mediums with different Nim oil (*Azadirachta indica*) concentrations. Ciencia Rural. 2004;**34**(6):1675-1680

[22] Depieri RA, Martinez SS, Ayres O, Menezes JR. Compatibility of the fungus *Beauveria bassiana* (Bals.) Vuill. (deuteromycetes) with extracts of neem seeds and leaves and the emulsible oil. Neotropical Entomology. 2005;**34**:601-606

[23] Mohan MC, Reddy NP, Devi UK, Ramesh K, Sharma HC. Growth and insect assays of *Beauveria bassiana* with neem to test their compatibility and synergism. Biocontrol Science and Technology. 2007;**17**(10):1059-1069

[24] Sahayaraj K, Namasivayam RKS, Rathi MJ. Compatibility of entomopathogenic fungi with extracts of plants and commercial botanicals. African Journal of Biotechnology. 2010;**10**(6): 933-938

[25] Halder J, Rai AB, Kodandaram MH. Compatibility of neem oil and different entomopathogens for the Management of Major Vegetable Sucking Pests. National Academy of Sciences. 2012; **36**(1):19-25

[26] Ribeiro LP, Blume E, Bogorni PC, Dequech STB, Brand SC, Junges E. Compatibility of *Beauveria bassiana* commercial isolate with botanical insecticides utilized in organic crops in southern Brazil. Biological Agriculture & Horticulture: An International Journal for Sustainable Production Systems. 2012;**28**(4):223-240

[27] Celar AF, Kos K. Compatibility of selected herbicides with entomo-pathogenic fungus *Beauveria bassiana* (Bals.) Vuill. Acta Agriculturae Slovenica. 2012;**99**(1):57-63. DOI: 10.2478/v10014-012-0007-2

[28] Gowrish RK, Ramesha B, Ushakumari R, Santhoshkumar T, Vijayaraghavakumar. Effect of spinosad 45 SC on growth and development of entomopathogenic fungi *Metarhizium anisopliae* and *Beauveria bassiana*. Entomon. 2013;**38**(3):155-160

[29] Challa MM, Sanivada KS. Compatibility of *Beauveria bassiana* (bals.) Vuill isolates with selected insecticides and fungicides at agriculture spray tank dose. Innovare Journal of Agricultural Science. 2014;**2**(3):7-10

[30] Faraji S, Shadmehri DA, Mehrvar A. Compatibility of entomopathogenic fungi *Beauveria bassiana* and *Metarhizium anisopliae* with some pesticides. Jounal of Entomological Society of Iran. 2016;**36**(2):137-146

[31] Joshi M, Gaur N, Pandey R. Compatibility of entomopathogenic fungi *Beauveria bassiana* and *Metarhizium anisopliae* with selective pesticides. Journal of Entomology and Zoology Studies. 2018; **6**(4):867-872

[32] Wisuda LN, Harjaka T, Prasetyaningrum DI. Compatibility between *Beauveria bassiana* (Bals.) and neem extract against Brown Plant-hopper (*Nilaparvata lugens* Stal.). Ilmu Pertanian (Agricultural Science). 2019; **4**(1):9-14

[33] Afandhi A, Pratiwi VR, Hadi MS, Setiawan Y, Puspitarini RD. Suitable combination between *Beauveria bassiana* (Balsamo) vuillemin and four plant leaf extracts to control *Spodoptera litura* (Fabricius). AGRIVITA Journal of Agricultural Science. 2020;**42**(2): 341-349 https://doi.org/10.17503/agrivita.v42i2.2678

[34] Celestino NF, Pratissoli D, Gonçalves dos Santos Junior JH, Machado CL. In vivo compatibility between *Beauveria bassiana* (Bals.) Vuillemin and castor oil on *Hypothenemus hampei* (Ferrari). *Coffee Science*. 2020;**15**(I):e151771. DOI: 10.25186/.v15i.1771. ISSN 1984-3909

[35] Pessoa AGL, Souza NMT, Loureiro de Souza E. Compatibility some insecticides used in pest management in eucalyptus with *Beauveria bassiana* (Cordycipitaceae). Research, Society and Development. 2020; **9**(8) e322985148 http://dx.doi.org/10.33448/rsd-v9i8.5148

[36] Hussain A. Compatibility of *Beauveria bassiana* and a plant secondary metabolite: A novel modeling approach to invade host defense for effective control of *Oligonychus afrasiaticus* (McGregor) on date palms. Journal of Fungi. 2021;7:334. DOI: 10.3390/jof7050334

[37] Manzanares BL, Villar EM, Mancebon VSM, Moreno IP. Compatibility of the entomopathogenic fungus *Beauveria bassiana* with etoxazole, spirodiclofen and spiromesifen against *Tetranychus urticae*. Biological Control. 2022;**169**(2022) 104892. DOI: 10.1016/j.biocontrol.2022.104892

[38] Stoll G. Protection naturelle des végétaux en zones tropicales.

AGRECOL-CTA. Josef Margraf ed. 97985 Weikersheim (Allemagne). 1988. p. 180

[39] Gata-Gonçalvès L. Propriedades Biocidas de Plantas de Origem Tropical: Caso da *Thevetia peruviana* Schum. PhD. Thesis. Engenharia Agronomica Quinina. Campos das Cebolas Lisboa; 2001 133 p.

[40] Dorothée MN, François ME, Nourou KNA, Angèle NP, Zachée A. Pathogenicity and in vitro control of *Lasiodiplodia theobromae* and *fusarium* sp., pathogens associated with cocoa dieback in Cameroon. International Journal of Current Research in Biosciences and Plant Biology. 2019;**6**(7):1-13

[41] Messi LM, Manga EF, Voula AV, Essouma MJE, Begoude BAD. In vitro study of the compatibility of six fungicides with two strains of *Trichoderma asperellum*, biocontrol agents used against cacao black pod disease in Cameroon. International Journal of Innovation and Applied Studies. 2018;**24**(4):1834-1848

[42] Razafindratiana E. Champignon entomopathogène et huiles de formulation: Test de formulation des spores de champignon entomopathogène dans les différents types d'huiles végétales existantes à Madagascar. GTZ, Documentation DEPSE; 2006. pp. 332-337

[43] Mamza WS, Zarafi AB, Alabi O. *In-vitro* evaluation of six fungicides on radial mycelia growth and re-growth of *fusarium pallidoroseum* isolated from Castor (*Ricinus communis*) in Samaru, Nigeria. African Journal of General Agriculture. 2008;**4**(2):65-71

[44] Oliveira RC, Neves PMOJ. Biological control compatibility of *Beauveria bassiana* with acaricides. Neotropical Entomology. 2004;**33**:353-358

[45] Opara EU, Wokocha RC. Efficacy of some plant extracts on the in-vitro and in-vivo control of *Xanthomonas campestris* Pv. Vesicatoria. Agricultural Journal. 2008;**3**(3):163-170

[46] Shapiro SS, Wilk MB. An analysis of variance test for normality (complete samples). Biometrika. 1965; **52**:591-611

[47] Fisher S, Yates. Statistical Tables for Biological Agricultural and Medical Research. 5ème ed. *Longman Group Limited*, London. 1963. p.146

[48] Jayarama K. Manuel de statistique pour la recherche forestière. Organisation des nations unies pour l'alimentation et l'agriculture (FAO); Rome (Italy). Dept. des Forets fre 185392; Commission Europeenne, Brussels (Belgium) fre 185393; 1999. p. 242

[49] Hassan SA. Testing methodology and the concept of the IOBC/WPRS working group. In: Jepson PC, editor. Pesticides and Non- Target Invertebrates. Dorset: Intercept, Wimborne; 1989. pp. 1-8

[50] Daumas F. Méthodes de normalisation des données. Revue de statistique appliquée, France. 1982;**3**(4): 23-38

[51] Woodill AJ, Nakamoto ST, Kawabata AM, Leung P. Optimal spraying strategy to combat the coffee berry borer: A dynamic approach. Journal of Agriculture and Food Research (JAFR). 2021;**4**:9. DOI: 10.1016/j. jafr.2021. 100125

[52] Johnson AM, Ruiz-Diaz PC, Manoukis CN, Jose CVR. Coffee berry borer (*Hypothenemus hampei*), a global Pest of coffee: Perspectives from historical and recent

invasions, and future priorities. Insects. 2020;**11**:882. DOI: 10.3390/insects111 20882

[53] Usha J, Naren BM, Padmaja V. Detection of compatibility of entomopathogenic fungus *Beauveria bassiana* (bals.) Vuill. With pesticides, fungicides and botanicals. IJPAES. 2014; **4**(2):613-624

[54] Deb L, Rajesh T, Majumdar D, Tombisana RK. Evaluation of biological compatibility of *Beauveria bassiana* with fungicides and botanicals. Journal of Pharmacognosy and Phytochemistry. 2017;**SP1**:1120-1124

[55] Dara SK. Compatibilité du champignon entomopathogène *Beauveria bassiana* avec certains fongicides utilisés dans la fraise en Californie. Open Plant Journal. 2019;**11**:7p

[56] MINAAPRA. Le biocontrôle pour la protection des cultures : 15 recommandations pour soutenir les technologies vertes. Rapport au Premier ministre François Fillon. France: Ministère de l'Agriculture, de l'Alimentation, de la Pêche, de la Ruralité et de l'Aménagement du territoire; 2011. p. 156

[57] Yoon M, Cha B, Kim J. Recent trends in studies on botanical fungicides in agriculture. The Plant Pathology Journal. 2013;**1**:1-9

[58] Alizadeh A, Samih MA, Khezri M, Riseh RS. Compatibility of *Beauveria bassiana* (Bals.) Vuill. With several pesticides. International Journal of Agriculture and Biology. 2007; **9**:31-34

[59] Bajan C, Kmitowa K, Popowska NE. Reaction of various ecotypes of entomopathogenic fungus *Beauveria bassiana* to the botanical preparation NEEMTM and pyrethroid Fastak. Archives of Phytopathology and Plant Protection. 1998;**3**:369-375

[60] Hiroze E, Neves JOMP, Zequi CAJ, Martins HL, Peralta HC, Alcides Moino Jr A. Effect of biofertilizers and neem oil on the entomopathogenic fungi *Beauveria bassiana* (Bals.). Vuill. and Metarhizium anisopliae (Metsch.) Sorok. Brazilian Archives of Biology and Technology. 2001;**44**(4): 419-423

[61] Dutta P, Bhattacharyya PN, Sarmah SR, Madhab M, Sandilya SP, Gogoi D, et al. *In vitro* studies on the compatibility assessment of certain agrochemicals with microbial biopesticides used in tea [*Camellia sinensis* (L.) O. Kuntze] of Assam, Northeast India. Two and a Bud. 2016; **63**(1):13-16

[62] Kakati N, Dutta P, Das P, Nath DP. Compatibility of Entomopathogenous fungi with commonly used insecticides for Management of Banana Aphid transmitting *Banana bunchy top virus* (BBTV) in Assam Banana production system. International Journal of Current Microbiology and Applied Sciences. 2018;**7**(11):2507-2513

[63] Rodriguez-Lagunes DA, Tajada AL, Diaz DR, Maciel CR, Mendoza JV, Roamn EB, et al. Compatibility of *Beauveria bassiana* and aqueous extracts of neem (*Azadirachta indica*) to control coffee borer (*Hypothenemus hampei*). Manejo Integrado de Plagas. 1997;**44**: 14-19

[64] Potrich M, Dallacort S, Luckman D, Lozano RE, Pegorini SC, Simionatto D. Compatibility of *Beauveria bassiana* and alternative phytosanitary products. Journal of Applied Microbiology. 2018:10

[65] Hernandez MM, Martinez-Villar E, Peace C, Perez-Moreno I, Marco V. Compatibility of the entomopathogenic fungus *Beauveria bassiana* with flufenoxuron and azadirachtin against Tetranychus urticae. Experimental and Applied Acarology. 2012;**58**:395-405

[66] Rehman H, Rasul A, Farooqi AM, Aslam UMH, Majeed B, Sagheer M, Ali Q. Compatibility of some botanicals and the entomopathogenic fungus, *Beauveria bassiana* (Bals.), against the red flour beetle, Tribolium castaneum (Herbst) (coleoptera: Tenebrionidae) Egyptian Journal of Biological Pest Control 2020;**30**(2020):131. https://doi.org/10.1186/s41938-020-00329-7

[67] Sajjad A, Muhammad AF, Sajjad A, Muhammad IU, Ahmad KQ, Siddique B, et al. Compatibility of entomopathogenic fungi and botanical extracts against the wheat aphid, *Sitobion avenae* (fab.) (Hemiptera: Aphididae). Egyptian Journal of Biological Pest Control. 2018; **28**(97):6p

[68] Duarte RT, Gonçalves KC, Espinosa DJL, Moreira LF, De Bortoli SA, Humber RA, et al. Potential of entomopathogenic fungi as biological control agents of diamondback moth (lepidoptera: Plutellidae) and compatibility with chemical insecticides. Journal of Economic Entomology. 2016;**109**(2): 594-601. DOI: 10.1093/jee/tow008

[69] Forim MR, Fernandes MFG, Cass QB, Fernandes JB, Vieira PC. Simultaneous quantification of azadirachtin and 3-tigloylazadirachtol in Brazilian seeds and oil of *Azadirachta indica*: Application to quality control and marketing. Analytical Methods. 2010;**2**: 860-869

[70] Shapiro-Ilan ID, Reilly CC, Hotchkiss WM, Wood WB. The potential for enhanced fungicide resistance in *Beauveria bassiana* through strain discovery and artificial selection. Journal of Invertebrate Pathology. 2002; **81**(2002):86-93

[71] Rodrigues Wenzel MI, Batista A, Giodano BI, Denadae EB, Fernandes BJ, Forim RM. Compatibility of polymers to fungi *Beauveria bassiana and Metarhizium anisopliae* and their formulated products stability. Acta Scientiarum Agronomy. 2016;**39**(4): 457-464. https://doi.org/10.4025/45 actasciagron.v39i4.32903.

[72] Islam TM, Omar BD. Combined effect of *Beauveria bassiana* with neem on virulence of insect in case of two application approaches. The Journal of Animal & Plant Sciences. 2012;**22**(1):77-82

[73] Gnanaprakasam AR, Sundaram J, Stephen DS, Kathirvelu B, Savariar V. Compatibility of entomopathogenic fungus *Beauveria bassiana* (Balsamo) Vuillemin isolated from Pulney hills, Western Ghats of Tamil Nadu with insecticides and fungicides. Elixir Agriculture. 2011;**40**(2011):5563-5567

[74] Cai Y, Zhang SL, Li ZZ. Compatibility of *Beauveria bassiana* to some Chemecal insecticides and herbicides. Chinese Journal of Biological Control. 2011;**27**(3):316-323

[75] Oliveira CN, Neves PMOJ, Kawazoe LS. Compatibility between the entomopathogenic fungus *Beauveria bassiana* and insecticides used in coffee plantations. Science in Agriculture. 2003;**60**(4):663-667

[76] Zacchino S, Lopez S, Galtusos MJ, De D, Gruañes A, Vivote GN, et al. In vitro antifungal evaluation and studies on mode of action of eight selected species from Argentine flora. Phytomédécin. 1998;**5**(5):339-395

Section 2

Pest Control through Biological Contol Agenet

Chapter 2

Perspective Chapter: Secondary Metabolites of Entomopathogens as Biotechnological Tools for the Biological Control of Agricultural Insect Pests

Hugo Rodríguez-Romero, Laura Rodríguez-Peláez, Adriana Reyes-Castro, Orlando Tlakaele Notario-Rendón, Manuel González-Peréz, Juventino Reza-Salgado, Pedro Antonio Rodríguez-Salazar and Jesús Antonio Salazar-Magallón

Abstract

In recent years, the irrational application of chemical insecticides has caused the appearance of pest insect populations that are resistant to the active principles of commercial insecticides. In addition, these chemical compounds cause significant damage to the environment and to the people who apply them. The use of secondary metabolites produced by entomopathogenic microorganisms is a viable alternative that could mitigate the damage caused by chemical insecticides. Actually, the secondary metabolites of entomopathogens microorganisms have been studied; however, there are few reports on their massive production and their direct application as biological control agents. The aim of this book chapter is to describe, in a very general way, some of the secondary metabolites produced by entomopathogenic microorganisms, their potential application as bioinsecticides as well as their mass production.

Keywords: secondary metabolites, beauvericin, cry proteins, destruxins, biological control

1. Introduction

The presence of pest insect populations resistant to some chemical insecticides, caused by indiscriminate and irrational use, has produced enormous agricultural losses throughout the world. For example, a solution to combat the increase of pest insect populations has been to increase the recommended doses and the application times of the chemical insecticide, with consequent damage to the environment [1].

On the other hand, the impact on the exports of agricultural products is due to the restrictions given in the European Union and the United States regarding residual chemicals in vegetables and fruits [2].

Actually, an alternative to the use of chemical insecticides, is the application of entomopathogenic fungi and bacteria. These microorganisms have been used as biological control agents for pest insects since the beginning of the last century. For example, in Mexico, the use of Entomopathogenic Microorganisms (EM) has not become widespread, although the application of these microorganisms as part of plant health began more than 50 years ago; however, a significant increase in the use and commercialization of biological products have been observed since 1990 in all the world [3].

In fact, the reason for the increase in the use and applications of EM throughout the world has been because of its efficiency in killing insect pests, remaining long in the field after application, in addition to its specific interaction with the insect pest and be relatively safe in terms of the environment. On the other hand, the mechanisms of pathogenicity of EM have also been extensively studied and some of the compounds, called secondary metabolites, that participate in infectious processes have been described. This knowledge has allowed establishment of strategies to improve the production of secondary metabolites and their application as biological control agents [4].

The secondary metabolites are a group of compounds that have a vital role in infective and control processes. These compounds synthesized by EM do not play a direct role in growth or reproduction but rather have an adaptation function to the environment which surrounds them. Furthermore, they have their origin as derivatives of various intermediate compounds in primary metabolism. EM secrete a wide range of secondary metabolites that can be used in biological control [5].

For example, the secondary metabolites synthesized by Entomopathogenic Fungi (EF), such as oxalic acid, beauvericins, and dextrixins, which are toxic against insects. These compounds are produced when the fungus has penetrated the exoskeleton and has reached the hemocele, that is when it is considered that have insecticidal properties [6, 7]. Equally, some bacteria can produce proteins with insecticidal capacity, as is the case with the bacterium *Bacillus thuringiensis*. These proteins called Cry proteins and are capable of controlling insects, such as Lepidoptera, Coleoptera, and Diptera, among others. Certain virulence factors, such as VIP proteins, S-layer proteins, and some enzymes with insecticidal capacity, have also been described [8].

The aim of this book chapter is to describe the potential of the main secondary metabolites produced by entomopathogenic microorganisms, as biological control agents, as well as analyze the main routes and bioprocesses of biotechnological production of these compounds.

2. Secondary metabolites of entomopathogenic fungi

Currently, many secondary metabolites generated by different entomopathogenic fungi have been reported. Some secondary metabolites may be of simple organic structure, but regularly they are compounds of a slightly more complex structure. Furthermore, many secondary metabolites are cyclic and linear peptide toxins, which are derived from primary metabolites, and in some cases with unusual structures and occasionally accompanied by processes of specific biosynthesis [9, 10].

2.1 Low molecular weight metabolites of entomopathogenic fungi

In recent research, a considerable number of low molecular weight secondary metabolites have been reported, these compounds have been isolated from insect pathogens. **Figure 1** shows some secondary metabolites with insecticidal activity produced by entomopathogenic fungi. These metabolites have simple structures, such as oxalic acid, 2,6-pyridinedicarboxylic acid (dipicolinic acid), 4-hydroxymethylazoxybenzene-4-carboxylic acid. Some reports describe that the secondary metabolites of entomopathogenic fungi can alter the permeability of insect cell membranes, inducing the loss of fluids in the cells, they also modify the molting and metamorphosis process, change in fertility, and interferes with interactions. Ligand-receptor occur in the plasmatic membrane, deformations in the wings, and finally, cause the death of the insect [11].

2.1.1 Oxalic acid

The production of this secondary metabolite with insecticidal activity has been reported in *Beauveria* sp., *Lecanicillium (Verticillium) lecanii*, *Paecilomyces fumosoroseus* and *Metarhizium anisopliae*. This substance has been described as a virulence factor in phytopathogenic fungi, in addition to being studied as an element that participates in the solubilization of cuticular protein [12]. There are some reports where oxalic acid has been used to control *Varroa destructor*; this compound has been applied to the colony using the spray technique (obsolete method due to its complicated handling), by dripping and sublimation. All techniques have been highly effective (90–95% or more) in broodless colonies and therefore meet the requirements for a winter treatment [13]. According to Ref. [14], a solution of oxalic acid and ethanol-water in a ratio of 0.05, 0.1, 0.2, and 0.4 w/v; can be administered in amounts of 200 μL per treatment; in a single application and therefore meets the needs of the large-scale beekeeping industry.

2.1.2 2,6-pyridinedicarboxylic acid

This important compound (dipicolinic acid) has been produced by some entomopathogenic fungi, among which *Isaria* sp., *M. anisopliae, B. bassiana*, and *L (V) lecanii* [15, 16]. Its sodium salt has insecticidal properties against *Spodoptera frugiperda* larvae [16].

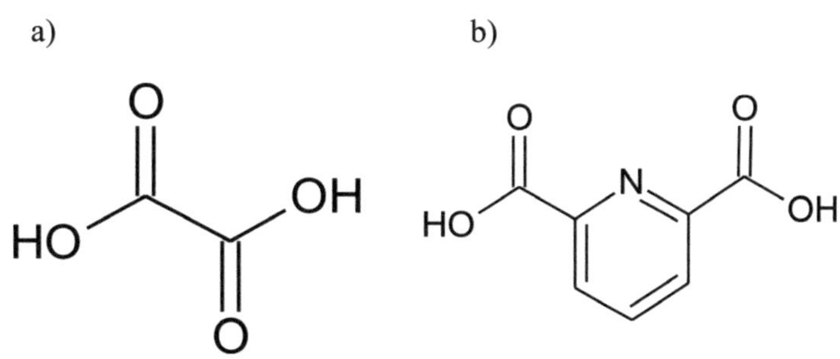

Figure 1.
Structures of low molecular weight secondary metabolites. (a) Oxalic acid produced by Beauveria spp., Lecanicillium (Verticillium) lecanii, Paecilomyces fumosoroseus, and Metarhizium anisopliae. (b) Dipicolinic acid produced by Isaria spp., M. anisopliae, B. bassiana, and L (V). Lecanii.

2.1.3 4-hydroxymethylazoxybenzene-4-carboxylic acid

It has been isolated together with its oxidation product (azoxybenzene-4,4-dicarboxylic acid) from the filtrate of the culture broth of *Entomophthora virulenta*, proving to be responsible for the insecticidal activity exhibited by this fungus against *Calliphora erythrocephala* [17].

2.2 Toxins with a peptide nature

Currently, cyclic and linear peptide toxins have been reported (**Figure 2**). The insecticidal action of these compounds has been described as very specific for certain groups of insects and their toxicity is due to the synergistic action of a complex group of compounds [18].

2.2.1 Beauvericin

Beauvericin has been the first molecule to be characterized due to its natural insecticidal properties, this compound was isolated for the first time from the mycelium of *Beauveria bassiana*, and in subsequent works, it has been extracted from different species of *Fusarium* and *Paecilomyces* species [18, 19]. Furthermore, beauvericin has been described as a cyclopeptide and its biosynthesis involves a multifunctional enzyme known as eniatin synthetase, whose expression is constitutive [18, 19]. In addition, it is known that this compound is an ionophore agent capable of forming Ca^+ and K^+ ion complexes that increase the natural and artificial permeability of membranes, inducing dehydration of tissues due to fluid loss from cells. Some reports indicate that this would be the main cause of the death of the insect, besides causing changes in the nucleus of the cells. This molecule has also been described and forms alterations in the molting and metamorphosis processes, as well as in fertility [19]. Some studies have reported that it has insecticidal properties against *Aedes aegypti* and *Hypothenemus hampei* mosquito larvae, producing paralysis after 6 hours of treatment and 73% mortality at 72 hours [20].

2.2.2 Efrapeptins

These molecules constitute a highly varied mixture of antibiotic peptides generated by some fungi, such as *Tolypocladium niveun*, *Beauveria nivea*, and *Tolypocladium*

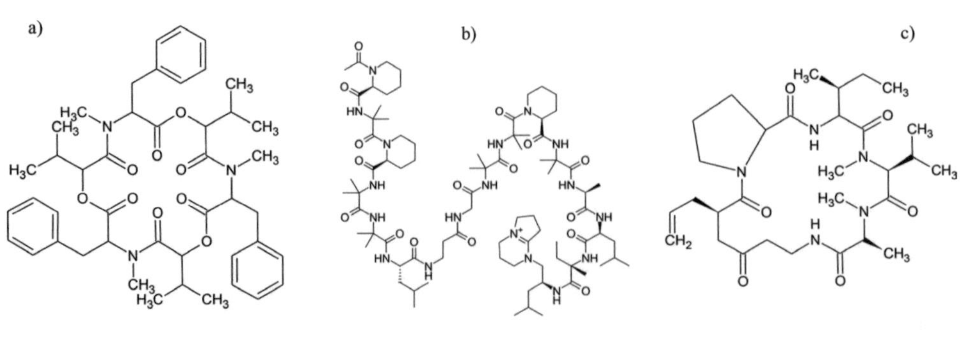

Figure 2.
Chemical structures of some toxins of a peptide nature. (a) Beauvericin, produced by Beauveria bassiana and Paecilomyces sp. (b) Efrapeptin F generated by Tolipocladium niveun, Beauveria nivea, and Tolipocladium cylindrosporum. (c) Destruxin a produced by Metarrizhium anisopliae.

cylindrosporum [21]. A nonpolar extract from a culture broth of *Beauveria bassiana* showed toxic activity against the red potato beetle. In fact, this extract, after appropriate fractionation, resulted in a peptide fraction identified as a mixture of efrapeptins, which was isolated by HPLC in five different peaks. These peaks were purified and characterized chromatographically, turning out to be rich in alpha-aminoisobutyric amino acid, and contained at least 15 other amino acid residues, in addition to having the acetylated N-terminus [22]. Efrapeptins produced by T. *cylindrosporum* have been shown to interfere with agglutinin in *Galleria mellonella* larvae, suggesting that these metabolites may interfere with ligand-receptor interactions that occur at the plasma membrane, allowing us to infer that most interactions between cells and humoral components of the insect immune system are receptor-mediated [23].

2.2.3 Destruxins

They are the best-characterized compounds since their mode of action also inhibits DNA, RNA, and protein synthesis in insect cells. One of the first studies aimed at the detection of fungal toxic substances has been on *Metarhizium anisopliae* strains, leading to the isolation of two cyclic depsipeptides that were called destruxins A and B. Today, 14 species of destruxins with variable insecticidal activities are known. This type of toxin has been analyzed and is capable of inhibiting the secretion of fluids through the Malpighi tube in *Schistocerca gregaria* [24]. The destruxins A, B, and E that have been produced by *M. anisopliae*, exhibited insecticidal properties when tested in *Plutella xylostella* larvae, with a high level of mortality in insect populations. They also caused deformations in the elytra and forewing of the insect. The influence of destruxin E on calcium flux and protein phosphorylation has also been observed *in vitro* cultures of Lepidoptera [25].

On the one hand, bassiacridin is a toxic protein, it has been purified from a strain of Beauveria bassiana by chromatographic methods. In a recent study, bassiacridin showed no affinity for anion exchangers and was characterized as a 60 KD monomer and an isoelectric point of 9.5. Furthermore, this molecule was shown to have β-glucosidase, N-galactosidase, and N-acetylglucosaminidase activity, as well as a proven insecticidal action [26].

3. *B. thuringiensis* secondary metabolites

δ-endotoxins produced particularly by *B. thuringiensis* (Bt) are parasporal protein bodies made up of polypeptide units of different molecular weights, from 27 to 140 kDa. Currently, 300 holotypes of *B. thuringiensis* toxins have been reported, classifying them into 73 Cry and 3 Cyt families. Bt toxins are produced during the sporulation phase, the Cry (crystal) protein is known for its specific toxic effects on a target organism (most belong to the order of insects), likewise the Cyt (cytolytic) proteins have been related with toxic effects on a wide variety of insects, mainly Diptera; however, its cytotoxicity against mammalian cells has also been proven [27].

Furthermore, during the vegetative cycle, *B. thuringiensis* produces soluble virulence factors of various types; some researchers have proposed that they be considered additional virulence factors to δ-endotoxins, which would explain the high pathogenicity of this bacterium toward its target organisms. In addition, some of these alternative virulence factors include -β-Exotoxins (Thuringiensins), nucleotide analogs of ATP and UTP that are synthesized by enzymes whose genes are found on conjugative plasmids, compete for RNA polymerases and thus block rRNA synthesis in mammals; by inhibiting transcription, they exert mutagenic activity. They are used for house fly control in Africa. VIP proteins (Vegetative Insecticidal Proteins), these proteins have crystallized and contain a domain similar to the active site of proteins with ADP-ribosylation activity SIP proteins [28].

3.1 Cry proteins

Cry toxins were the first described proteins in *B. thuringiensis*, being the most diverse group, with 78 different protein families and a total of 817 described proteins. Cry proteins are active against a wide range of insect species (Lepidoptera, Coleoptera, Aphids, Nematodes, etc.). The continuous and growing discovery of a wide variety of Cry toxins then demanded the creation of a nomenclature system that would allow them to be classified in an orderly manner. The classification system for these toxins was designed and proposed by the *B. thuringiensis* Toxin Nomenclature Committee, which was based on the identity percentage of the amino acid sequence of each protein. In this way, when a new *B. thuringiensis* toxin is discovered, it is assigned a name by comparing it with its closest counterpart [29].

Additionally, the mechanism of action of Cry proteins was mainly described in Lepidoptera as a multistep process. *B. thuringiensis* crystals are ingested and then solubilized in the insect's midgut, after which the crystal proteins are released as protoxins. These will not cause damage *per se* but must be processed by intestinal proteases to generate active toxins that will lead to the death of the larva. In their monomeric form, the toxins cross the peritrophic membrane and bind univalent to cadherin, with great affinity for the apical face of the epithelial membrane. Then, according to studies carried out in insect cell cultures, a magnesium ion-dependent signaling cascade is initiated that would be responsible for cell death [30].

Furthermore, the initiation of this signaling cascade stimulates the exocytosis of cadherin from intracellular vesicles to the apical membrane of the cell and increases the number of receptors; therefore, it recruits a greater number of free toxins that would amplify the initial signal. On the other hand, based on *in vitro* experiments and bioassays, it is proposed that the binding of monomers to cadherin facilitates proteolytic cleavage on the N-terminus of the toxin. This last cleavage induces the assembly of the monomers and a pre-pore oligomeric form is established. This structure increases the affinity for a secondary receptor, such as aminopeptidase N or alkaline phosphatase. Finally, binding to this second receptor facilitates the formation of a pore in the epithelium of the midgut, which causes an osmotic imbalance and the consequent cell lysis. The intestinal tissue is seriously damaged, which prevents the assimilation and retention of vital compounds for the larva and leads to the death of the insect. Death can be accelerated by germinating spores and proliferating vegetative cells in the insect's hemocoel [31].

Due to the success in biological control, some brands have developed commercial products based on *B. thuringiensis*. For example, The Abbott Company has, among others, products based on the kurstaki serovars like Dipel (cry1Aa, cry1Ab1, cry1Ac1, cry2Aa1, cry2Ab1 genes) and Xentari (*B. thuringiensis* aizawai with cry1Aa1, cry1Ab1, cry1Ba1, cry1Ca1, cry1Da1 genes) that they are lethal against lepidopteran insects. Ecogen Inc. has promoted the biopesticides Lepinox (based on *B. thuringiensis* kurstaki) and Crymax (*B. thuringiensis* strain ED7826 with cry1Ac, cry2A, and cry1C genes) for lepidopteran lethality. Biochem has developed a formula based on *B. thuringiensis* israelensis for the control of Diptera (Cry4Aa1, Cry4Ba1,

Cry10Aa1, Cry11Aa1 toxins). Mycogen developed M-trak, a product for which it was sued by the company Novo Nordisk for misuse of the strain [32].

3.2 Cyt proteins

Cyt proteins have been found in the parasporal crystal produced by *B. thuringiensis* and they are active against Diptera. These proteins present cytolytic activity *in vitro* assays so that Cyt proteins break intestine cells due to their cytolytic activity and not by forming pores like Cry proteins. So far, known members of the Cyt proteins include the Cyt1, Cyt2, and Cyt3 protein families. Cyt proteins have a molecular weight of 28 kDa, which is a cytolysin of nonspecific action, produced mainly by the israelensis variety, accumulating in the crystal, together with the endotoxins typical of this variety. Because they have no homology to the other Cry proteins, the Cyt toxins are not classified as endotoxins. The general structure of Cyt proteins has an α/β-like domain with a β-sheet in the center surrounded by two α-helicals. The central β-sheet consists of six antiparallel β-strands, flanked by a layer of α-helices [33].

According to the studies of Ref. [34], the amplification of fragments of the expected size was obtained by PCR using pairs of oligonucleotide primers that detect the genes cry4Aa, cry4Ba, cry11Aa, cry11Ba, cyt1Aa, cyt1Ab, and cyt2Aa. The analyzes confirmed that, of the 1073 isolates subjected to PCR, only 45 (4.2% of the total) presented amplification for a single gene or the combination of them among some isolates. Of the 45 isolates (specific to Diptera) of *B. thuringiensis* subjected to selective bioassays, 13 (28.9%) presented 100% mortality against *A. aegypti* larvae, which were subjected to quantitative bioassays for the calculation of LC_{50} and LC_{90}.

The relationship between toxicity and gene content of the *B. thuringiensis* isolates studied suggests a joint action of the toxins in the control of *A. aegypti*. Isolates carrying three or more Diptera-specific genes (cry and cyt) were effective in selective bioassay tests, causing 100% mortality, which is equivalent to treatment with the standard strain *B. thuringiensis* var. israelensis. Some isolates showed three combinations of genes and showed no control efficacy against insect larvae. In these combinations, the cry4Aa and cry4Ba genes were not present and, in isolates 574 and 582, the cry11Aa gene was not present either [33, 34].

In studies carried out by [35], the strains LBT-63 and LBT-87 have shown amorphous crystals very similar to those found in *B. thuringiensis* subs. Israelensis; however, strains LBT-62 and LBT-99 showed very small inclusions attached to the spores. The evaluation of the biological activity of strains LBT −62 and LBT 83 and LBT-99 against *D. melanogaster* have shown how the LBT-87 strain has a 70% mortality in the larvae. Of the larvae that managed to survive and form pupae, they never reached the adult stage. The remaining strains only slightly delayed pupal formation, but adult emergence eventually occurred. Although the LBT-87 strain showed an effect against *D. melanogaster*, the virulence was lower than that obtained with the standard strain *B. thuringiensis* subs. Israelensis, specific for Diptera.

3.3 VIP proteins

VIP proteins (Vegetative Insecticidal Proteins) constitute another family of insecticidal proteins produced by some strains of *B. thuringiensis*. Such proteins do not resemble Cry proteins and are produced during the vegetative phase of the bacterium, in addition to being soluble, do not form crystals and have the same magnitude of toxicity as Cry proteins. In addition, the insecticidal spectrum of VIP proteins includes

Coleoptera (VIP1 and VIP2) and Lepidoptera (Vip3), including some species of insect pests that are insensitive to the action of Cry proteins. The *Vip2A(a) and Vip1A(a)* genes of some *B. thuringiensis* strains are known to be located in plasmids. Also, the proteins encoded by these genes contain a signal peptide typical of secretory proteins in the N-terminal region. In addition, it is known that both Cry and VIP proteins, toxic to beetle larvae, have similar sequences and probably a similar mode of action, since both are based on binding to specific membrane receptors of midgut epithelial cells of the target insects and in the consequent formation of pores [36].

Through histopathological observations, it was possible to verify that the epithelial cells of the midgut of susceptible insects are the main target of the insecticidal protein VIP3A, which causes intestinal paralysis, complete cell lysis, and the consequent death of the larvae. Thus, disruption of intestinal cells appears to be the main mechanism for the lethality of VIP proteins. In addition, the production of the VIP3A protein by vegetative cells after spore germination is an important factor in combined spore toxicity in insect species in which the Cry proteins are relatively inactive [37].

3.4 S-layer proteins

The surface layers (S-layers) proteins are known as monomolecular crystalline arrays of proteinaceous subunits, these compounds are over the surfaces of many bacteria and archaea [38]. Besides, S-layer proteins frequently have demonstrated a big capacity for self-assembly during their formative stage. Several signal peptides and three S-layer homology (SLH) domains, which are anchored to the cell surface are found at the N-terminal part of S-layer proteins and various S-layer genes from *B. thuringiensis* have been cloned and sequenced [39]. Among these *B. thuringiensis* strains, only one S-layer of *B. thuringiensis* subsp. finitimus strain CTC has been thoroughly studied for its biochemical characterization [40].

According to Ref. [41], four *B. thuringiensis* strains whose parasporal inclusions contained the S-layer protein (SLP) and cloned two slp genes from each strain. So, phylogenetic analysis indicated these SLPs could be divided into two groups, SLP1s and SLP2s. To confirm whether SLPs were present in the S-layer or as a parasporal inclusion, strains CTC and BMB1152 were chosen for further study. Western blots with isolated S-layer proteins from strains CTC and BMB1152 in the vegetative phase showed that SLP1s and SLP2s were constituents of the S-layer.

On the other hand, Ref. [42] reported the identification of an S-layer protein by the screening of *B. thuringiensis* strains for activity against various insect pests as the coleopteran *Epilachna varivestis* (Mexican bean beetle; Coleoptera: Coccinellidae). Some of the *B. thuringiensis* strains assayed against *E. varivestis* showed moderate toxicity. However, a *B. thuringiensis* strain (GP1) that was isolated from a dead insect showed remarkably high insecticidal activity. The parasporal crystal produced by the GP1 strain was purified and shown to have insecticidal activity against *E. varivestis* but not against the lepidopteran *Manduca sexta* or *Spodoptera frugiperda* or the dipteran *A. aegypti*.

4. Development and escalation in the production of secondary metabolites

In general lines, the development of biopesticides has a close relationship with the study and exploitation of modern biotechnology, therefore, so that the entities related to such initiatives (industry, universities, and research centers, among others) can generate innovation and technological development requires a highly qualified human resource of a multidisciplinary nature [43]. In this context, technological development can be understood as the use of existing scientific knowledge for the production of new materials, devices, products, procedures, systems, or services, as well as for their substantial improvement, which includes prototyping and pilot installations [44]. However, the technological development of a microbial biopesticide is a complex task that involves not only technical-scientific stages but also stages associated with the analysis of economic viability and market potential, as well as compliance with national regulations and regulations for bioproducts [45].

Other important aspects in the development of a biopesticide are the costs of implementation and execution, the relative complexity, and the duration of each stage. Samada et al. [46] have described as technological development progresses, possible microorganisms or candidate isolates are screened, but, in turn, the costs, complexity, and time required increase due to multiple tests, bioassays, analysis, and studies that must be carried out to ensure the efficiency and reproducibility of the bioproduct under controlled, semi-controlled, and field conditions. Once all the stages of the development of a biopesticide have been executed (which can take between 4 and 5 years), it can be ensured that the technological development is efficient, effective, and stable, which allows projecting the commercialization and distribution of the biopesticide in the emerging market of bioproducts worldwide.

The need for sustainable development of agriculture has fostered many initiatives that have stimulated the development of alternative methods that reduce the use of chemical pesticides in pest control. Microbial biopesticides, therefore, represent one of the most promising alternatives and, although their commercialization remains marginal, the demand is constantly increasing in all parts of the world. This significant increase coincides with the growth of biological pest control in high-value crops, such as greenhouse-produced vegetables, fruit crops, vineyards, and forestry, among others. On the other hand, despite the fact that biological control has shown an important development in organic agriculture, the most promising future of biopesticides is found in integrated pest management (IPM) programs [47].

Even though there are significant advances in scientific and technological knowledge on the development of biopesticides, there are few cases in which formulations with a high and consistent biocontrol activity have been produced, which also have a wide spectrum of use and respond to the challenges cheaply. Although many bioproducts have been developed, several of them have been withdrawn from the market before achieving commercial success. In recent years, there have been important advances in the development and industrial production of biopesticides, but they still occupy a small percentage of the products used for crop protection. In many cases, research centers or companies develop excellent biopesticides but fail to position them before producers. For this reason, a product is only effective if it generates impact in field conditions and if it is supported by a solid market strategy that guarantees reliability and profitability [48].

5. Conclusions and perspectives

The production of a biopesticide not only requires detailed knowledge of the microbiology and physiology of the biocontrol microorganism but also knowledge of the biology of the pest, the epidemiological aspects that define its harmful effect on the crop, as well as the physiology of the pest. Plant. In addition to this, several technological challenges related to the fermentation process, the type of formulation, the pest population versus the biocontrol microorganism population, and the application systems used must be addressed. Therefore, before the development and application of a formulation, it is necessary to understand the ecology of the interaction between the biocontrol microorganism, the host plant, and the pest. Likewise, in the development of biopesticides, each step must be considered: the selection of the microorganism, the production method, the delivery system, the application technology, the factors that affect its development and persistence in the environment and, ultimately, instance, the availability of the product in the market and the various positioning actions of this to achieve recognition and acceptance by farmers.

Acknowledgements

We thank the Faculty of Sustainable and Protected Agriculture, as well as the academic team in sustainable and protected agriculture, for the support provided during the completion of this study.

Conflict of interest

The authors declare no conflict of interest.

Author details

Hugo Rodríguez-Romero, Laura Rodríguez-Peláez, Adriana Reyes-Castro, Orlando Tlakaele Notario-Rendón, Manuel González-Peréz, Juventino Reza-Salgado, Pedro Antonio Rodríguez-Salazar and Jesús Antonio Salazar-Magallón*
Faculty of Engineering in Sustainable and Protected Agriculture, Agricultural Bioengineering Laboratory, Avenida Universidad Tecnológica #1, México

*Address all correspondence to: biologicalcontrollab@gmail.com

References

[1] Barratt BIP, Moran VC, Bigler F, et al. The status of biological control and recommendations for improving uptake for the future. BioControl. 2018;**63**:155-167. DOI: 10.1007/s10526-017-9831-y

[2] Rusch A, Valantin-Morison M, Sarthou JP, Roger-Estrada J. Biological control of insect pests in agroecosystems: Effects of crop management, farming systems, and Seminatural habitats at the landscape scale: A review. Advances in Agronomy. 2010;**109**:219-259. DOI: 10.1016/B978-0-12-385040-9.00006-2

[3] Brunner-MendozaC, Reyes-MontesMC, Moonjely S, Bidochka MJ, Torrielo C. A review on the genus *Metarhizium* as an entomopathogenic microbial biocontrol agent with emphasis on its use and utility in México. Biocontrol Science and Technology. 2018;**29**(1):83-102. DOI: 10.1080/09583157.2018.1531111

[4] Deka B, Baruah C, Babu A. Entomopathogenic microorganisms: Their role in insect pest management. Egyptian journal of biological. Pest Control. 2021;**31**:121. DOI: 10.1186/s41938-021-00466-7

[5] Litwin A, Nowak M, Rozalska S. Entomopathogenic fungi: Unconventional applications. Reviews in Environmental Science and Biotechnology. 2020;**19**:23-42. DOI: 10.1007/s11157-020-09525-1

[6] Lozano-Tovar MD, Garrido-Jurado I, Quesada-Moraga E, Raya-Ortega MC, Trapero-Casas A. *Metarhizium brunneum* and *Beauveria bassiana* release secondary metabolites with antagonistic activity against *Verticillium dahliae* and *Phytophthora megasperma* olive pathogens. Crop Protection. 2017;**100**:186-195. DOI: 10.1016/j.cropro.2017.06.026

[7] Weng Q, Zhang X, Chen W, Hu Q. Secondary metabolites and the risks of *Isaria fumosorosea* and *Isaria farinose*. Molecules. 2019;**24**(664):1-17. DOI: 10.3390/molecules24040664

[8] Sansinenea E, Ortiz A. Secondary metabolites of soil *Bacillus* spp. Biotechnology Letters. 2011;**33**:1523-1538. DOI: 10.1007/s10529-011-0617-5

[9] Sandhu SS, Shukla H, Aharwal RP, Kumar S, Shukla S. Efficacy of entomopathogenic fungi as green pesticides: Current and future prospects. In: Microorganisms for Green Revolution, Microorganisms for Sustainability. Springer Nature Singapore Pte Ltd; 2017;**6**:327-348. DOI: 10.1007/978-981-10-6241-4_17

[10] Gibson DM, Donzelli BGG, Krasnoff SB, Keyhani NO. Discovering the secondary metabolite potential encoded within entomopathogenic fungi. Natural Products Reports. 2014;**31**(10):1287-1305. DOI: 10.1039/C4NP00054D

[11] Sanchéz-Pérez LC, Barranco-Florido JE, Rodríguez-Navarro S, Cervantes-MayagoitiaJF, Ramos-LópezMA. Enzymes of entomopathogenic fungi, advances and insights. Advance in Enzymes Research. 2014;**2**:65-76. DOI: 10.4236/aer.2014.22007

[12] Khachatourians GG, Qazi SS. Entomopathogenic fungi: Biochemistry and molecular biology. In: Brakhage AA, Zipfel PF, editors. Human and Animal Relationships, the Mycota VI. 2nd ed. Berlin Heidelberg: © Springer-Verlag; 2008. pp. 1-30. DOI: 10.1007/978-3-540-79307-6_3

[13] Kirkland BH, Eisa A, Keyhani NO. Oxalic acid as a fungal acaracidal virulence factor. Journal of Medical Entomology. 2005;**42**(3):346-351. DOI: 10.1093/jmedent/42.3.346

[14] Maggi MD, Damiani N, Ruffinengo SR, Brasesco MC, Szawarski N, Mitton G, et al. The susceptibility of *Varroa destructor* against oxalic acid: A study case. Bulletin of Insectology. 2017;**70**(1):39-44

[15] Watanabe N, Hattori M, Yokohama E, Isomura S, Ujita M, Hara A. Entomogenous fungi that produce 2,6-pyridine dicarboxylic acid (dipicolinic acid). Journal of Bioscience and Bioengineering. 2006;**102**(4):365-368. DOI: 10.1263/jbb.102.365

[16] Han JH, Jin BR, Kim JJ, Lee SY. Virulence of entomopathogenic fungi *Metarhizium anisopliae* and *Paecilomyces fumosoroseus* for the microbial control of *Spodoptera exigua*. Mycobiology. 2014;**42**(4):385-390. DOI: 10.5941/MYCO.2014.42.4.385

[17] Subbanna AR, Stanley J, Rajasekhara H, Pattayanak A, Bhowmick R. Perspectives of microbial metabolites as pesticides in agricultural pest management. In: Mérillon JM, Ramawat K, editors. Co-Evolution of Secondary Metabolites. Reference Series in Phytochemistry. Cham: Springer; 2017. DOI: 10.1007/978-3-319-96397-6_44

[18] Wang Q, Xu L. Beauvericin, a bioactive compound produced by fungi: A short review. Molecules. 2012;**17**:2367-2377. DOI: 10.3390/molecules17032367

[19] Mallebrera B, Prosperini A, Font G, Ruiz MJ. *In vitro* mechanisms of Beauvericin toxicity: A review. Food and Chemical Toxicology. 2018;**111**:537-545. DOI: 10.1016/j.fct.2017.11.019

[20] Daniel J, Silva A, Nakagawa D, de Madeiros L, Carvalho M, Tavares L, et al. Larvicidal activity of *Beauveria bassiana* extracts against *Aedes aegypti* and identification of beauvericins. Journal of the Brazilian Chemical Society. 2017;**28**(6):1003-1013. DOI: 10.21577/0103-5053.20160253

[21] Brückner H, Degenkolb T. Sequences of tolypins, insecticidal efrapeptin-type peptaibiotics from species of the fungal genus *Tolypocladium*. Chemistry and Biodiversity. 2020;**17**:e200027. DOI: 10.1002/cbdv.202000276

[22] Islam W, Adnan M, Shabbir A, Naveed H, Abubakar YS, Qasin M, et al. Insect-fungal-interactions: A detailed review on entomopathogenic fungi pathogenicity to combat insect pests. Microbial Pathogenesis. 2021;**159**:105122. DOI: 10.1016/j.micpath.2021.105122

[23] Bandani AR. Effects of *Tolypocladium cylindrosporum* and its secondary metabolites, efrapeptins, on the immune system of *galleria mellonella* larvae. Biocontrol Science and Technology. 2005;**15**(1):67-79. DOI: 10.1080/09583150400015961

[24] Skrobek A, Shah FA, Butt TM. Destruxin production by the entomogenous fungus *Metarhizium anisopliae* in insects and factors influencing their degradation. BioControl. 2008;**53**:361-373. DOI: 10.1007/s10526-007-9077-1

[25] Tiang-Yan D, Bo-Wen Z, Qun-Fang W, Qiong-Bo H. The production relationship of destruxins and blastospores of *Metarhizium anisopliae* with virulence against *Plutella xylostella*. Journal of Integrative Agriculture. 2016;**15**(6):1313-1320. DOI: 10.1016/S2095-3119(15)61322-3

[26] García-Estrada C, Cat E, Santamarta I. *Beauveria bassiana* as

biocontrol agent: Formulation and commercialization for Pest management. In: Singh H, Sarma B, Keswani C, editors. AgriculturallyImportantMicroorganisms. Singapore: Springer; 2016. DOI: 10.1007/978-981-10-2576-1_5

[27] Jouzani GS, Valijanian E, SharafiR.*Bacillusthuringiensis*:Asuccessful insecticide with new environmental features and tidings. Applied Microbiology and Biotechnology. 2017;**101**:2691-2271. DOI: 10.1007/s00253-017-8175-y

[28] Chandrasekaran R, Revathi K, Senthil-Nathan S, Kalaivani K, Hunter WD, Duraipandiyan V, et al. Eco-friendly formulation of wild *Bacillus thuringiensis* secondary metabolites through molecular characterization against the lepidopteran pest. Physiologycal and Molecular Plant Pathology. 2018;**101**:93-104. DOI: 10.1016/j.pmpp.2017.09.002

[29] Deng C, Peng Q, Song F, Dereclus D. Regulation of *cry* gene expression in *bacillus thuringiensis*. Toxins. 2014;**6**(7):2194-2209. DOI: 10.3390/toxins6072194

[30] De Almeida-Melo AL, Soccol TV, Soccol RC. *Bacillus thuringiensis*: Mechanism of action, resistance, and new applications: A review. Critical Reviews in Biotechnology. 2016;**36**(2):317-326. DOI: 10.3109/07388551.2014.960793

[31] Vachon V, Laprade R, Schwartz JL. Current models of the mode of action of *bacillus thuringiensis* insecticidal crystal proteins: A critical review. Journal of Invertebrate Pathology. 2012;**111**(1):1-12. DOI: 10.1016/j.jip.2012.05.001

[32] Bravo A, Gómez I, Porta H, García-Gómez BI, Rodríguez-Almazán C, Pardo L, et al. Evolution of *Bacillus thuringiensis* crytoxins insecticidal activity. Microbial Biotechnology. 2013;**6**(1):17-26. DOI: 10.1111/j.1751-7915.2012.00342.x

[33] Chengchen X, Bi-Cheng W, Ziniu Y, Ming S. Structural insights into *Bacillus thuringiensis* cry. Cyt and Parasporin. Toxins. 2014;**6**:2732-2770. DOI: 10.3390/toxins6092732

[34] Da Silva J, Rossi J, Marucci S, Alves E, Volpe H, Ferraudo A, et al. Toxic activity of *Bacillus thuringiensis* isolates to *Aedes aegypti* (L.) Larvae (Diptera: Culicidae). Neotropical Entomology. 2010;**39**(5):757-766

[35] Carreras-Solís B, Rodríguez-Batista D. Evaluación de cepas de *Bacillus thuringiensis* para el control de *Drosophila melanogaster* maengi. Fitosanidad. 2009;**13**(2):83-87

[36] Gupta M, Kumar H, Kaur S. Vegetative insecticidal protein (Vip): A potential contender from *bacillus thuringiensis* for efficient management of various detrimental agricultural pests. Frontiers in Microbiology. 2021;**12**. DOI: 10.3389/fmicb.2021.659736

[37] Chakrabarti S, Jin M, Wu C, Chakraborti P, Xiao Y. *Bacillus thuringiensis* vegetative insecticidal protein family Vip3A and mode of action against pest Lepidoptera. Pest Management Science. 2020;**76**(5):1612-1617. DOI: 10.1002/ps.5804

[38] Dong Z, Li J, Zheng J, Geng C, Peng D, Sun M. Complete genome sequence of *Bacillus thuringiensis* CTC—A typical strain with high production of S-layer proteins. Journal of Biotechnology. 2016;**220**:100-101. DOI: 10.1016/j.jbiotec.2015.12.027

[39] Wang L, Sun M, Yu Z. Capacity of Bacillus thuringiensis S-layer protein

displaying polyhistidine peptides on the cell surface. Applied Biochemestry and Biotechnology. 2004;**119**:133-143. DOI: 10.1385/ABAB:119:2:133

[40] Zhu C, Yu Z. The surface layer protein of Bacillus thuringiensis CTC forms unique intracellular parasporal inclusion body. Journal of Basic Microbiology. 2008;**48**(4):302-307

[41] Zhou Z, Peng D, Zheng J, Gou G, Tian L, Yu Z, et al. Two groups of S-layer proteins, SLP1s and SLP2s, in *Bacillus thuringiensis* co-exist in the S-layer and in parasporal inclusions. BMB Reports. 2011;**44**(5):323-328

[42] Peña G, Miranda-Ríos J, de la Riva G, Pardo-López L, Soberón M, Bravo A. A *Bacillus thuringiensis* S-layer protein involved in toxicity against *Epilachna varivestis* (Coleoptera: Coccinellidae). Applied and Environmental Microbiology. 2006;**72**(1):352-360. DOI: 10.1128/AEM.72.1.353-360.2006

[43] Muñiz-Paredes F, Miranda-Hernández F, Loera O. Production of conidia by entomopathogenic fungi: From inoculants to final quality tests. World Journal of Microbiology and Biotechnology. 2017;**33**:57. DOI: 10.1007/s11274-017-2229-2

[44] Koul O. Microbial biopesticides: Opportunities and challenges. Perspectives in Agriculture, Veterinary Science, Nutrition Resourses. 2011;**6**(56):1-26. DOI: 10.1079/PAVSNNR201160

[45] Montesinos E. Development, registration and commercialization of microbial pesticides for plant protection. International Microbiology. 2003;**6**:245-252. DOI: 10.1007/s10123-003-0144-x

[46] Samada L, Tambunan U. Biopesticides as promising alternatives to chemical pesticides: A review of their current and future status. Journal of Biological Sciences. 2020;**20**(2):66-76. DOI: 10.3844/ojbsci.2020.66.76

[47] Jallouli W, Driss F, Fillaudeau L, Rouis S. Review on biopesticide production by *bacillus thuringiensis* subsp. *kurstaki* since 1990: Focus on bioprocess parameters. Process Biochemestry. 2020;**98**:224-232

[48] El-Bendary M. *Bacillus thuringiensis* and *Bacillus sphaericus* biopesticides production. Journal of Basic Microbiology. 2006;**46**(2):158-170. DOI: 10.1002/jobm.200510585

Chapter 3

Perspective Chapter: Pheromone-Based Techniques in Sustainable Pest Management

Melis Yalçın

Abstract

The intensive use of insecticides can result in environmental pollution and adverse effects on human health due to the issue of insecticide residue in the environment. To mitigate this, various control techniques, including cultural, biological, and biotechnical methods, or their combinations, can be employed to manage invasive species. One such biotechnical method that has become popular is the use of pheromones. Pheromone techniques enable early pest detection, population monitoring, mass trapping or annihilation, and mating disruption. To maximize their effectiveness, it is important to determine the exact pheromone component, optimization rate, trap design, and saturation rate for each species. In conclusion, implementing these different pheromone-based strategies is essential for providing effective pest management strategies that take regional variations in pheromones into account.

Keywords: pheromone, pests, monitoring, mating disruption, pest management

1. Introduction

Pheromones that represent intraspecific communication of the species are considered environmentally friendly for monitoring and controlling insect species. Chemical communication leads to emitting species-specific signals of individuals which causes reactions in different life forms of the same species. In the past pheromones were described as; chemicals involved in intraspecific communication which describes the term "ectohormone" [1] but the term changed into pheromone in 1959 [2, 3]. Detection of the first sex pheromone of the silkworm moth, *Bombyx mori* (Lepidoptera: Bombycidae) has revealed the necessity of studies on pheromones [4]. In the last decades, more than 600 species of Lepidopteran pheromones were discovered [5, 6] as a consequence sex pheromone global market size increased to 2.4 billion USD per year in 2019 [7]. After World War II, the usage of broad-spectrum insecticides increased dramatically, leading to intensive usage of chemicals which in turn caused residues on food, development of pest resistance, environmental pollution, and posed risks to the health of humans and other living organisms [8]. In the past decades, there are lots of studies about pheromone biosynthesis, new insect pheromones, their modes of action, and their application in integrated pest management.

IntechOpen

Integrated pest management approach combined with chemical ecology, insect behavior and interaction between organisms provides exploring the pheromones of pests [9]. Pheromones are successful when the pest population density is low, they provide long-term reduction in pest populations and generally do not affect natural enemies.

Insect sex pheromones comprise different components usually one chemical is primary component and efficient to attract mates [10, 11]. Based on their chemical structure and biosynthetic origin, various types of pheromones exist. Type I pheromones, for instance, are characterized by the presence of acetates, alcohols, and aldehydes, and constitute approximately 75% of all moth sex pheromones that are currently known [12]. Type II pheromones consist of hydrocarbons and epoxide derivatives with carbon atoms [12] and contain about 15% of moth sex pheromones reported. Type III pheromones contain hydrocarbons that involve methyl branches. Type 0 pheromones involve methyl-carbinols and methyl ketones and generally belong to the Eriocraniidae and caddisflies (Tricoptera) [6, 13]. The chemical structure of pheromones comprises of carboxylic acids, hydrocarbons, epoxides, lactones, alcohols, esters, aldehydes, ketones, isoprenoids, and triacyl glycerides [6, 14, 15].

There are different kinds of pheromones in the nature. Releaser pheromones such as alarm pheromones cause sudden changes in the behavior of pests while primer pheromones cause slower and longer physiological changes [16–18]. In the presence of the menace, the secretion begins between the alert members of the same species. After sex pheromones, the alarm pheromones are the most common pheromone produced by insects. Generally, alarm pheromones of some aphid species involve germacrene A, α-pinene, and sesquiterpene (E)-β-farnesene (EBF) [19, 20]. Alarm pheromones are released by some insects to avoid attack by natural enemies. Avoidance and dispersal behavior can be observed when alarm pheromones are released. But some social insects such as bees and ants may respond aggressively to alarm pheromones [16]. Sex pheromones generally attract males over long distances and very low concentrations of sex pheromones can be detected by sensilla of insects [21]. Studies concerned about sex pheromones consist of Lepidoptera, Diptera, Hemiptera, Coleoptera, and Blattodea since 2000. Especially Noctuidae, Tortricidae, Plutellidae, Crambidae, Bombycidae, Pyralidae, Lymantriidae, Gelechiidae, Geometridae, Lasiocampidae, Sesiidae, Gracillaridae, Erebidae family from Lepidoptera order, Drosophilidae, Tephritidae, Psychodidae from Diptera order, Pseudococcidae and Miridae from Hemiptera order, Scarabaeidae and Buprestidae from Coleoptera order, Blatteidae from Blattodea order were the most studied ones. EPA gave permission in 1978 for the registration of Gossyplure which was the first registered product used for the management of pink bollworm [22]. Insect sex pheromones generally consist of more than three double bonds, aliphatics of 9–18 carbons, and end with an acetate, aldehyde, or alcohol [20]. The weather conditions, time, and host plant availability are important for the release of sex pheromones [17]. Aggregation pheromones provide to generate groups and mating [23] and some species of bark beetles (Scoliytidae: Coleoptera) release aggregation pheromones for mating and feeding [24]. Conversely, anti-aggregation pheromones cause dispersal of individuals for both sexes to find optimum space to feed. Oviposition-deterring pheromones warn insect species to avoid egg-laying on hosts such as females of fruit flies (Diptera: Tephritidae) *Ceratitis capitata* release oviposition determining fruit making pheromone after egg-laying [25]. Trail pheromones provide insects to find nest sites and to mark feeding for social insects such as termites and ants [18]. Home recognition pheromones are observed in social insect colonies. Queens of social insects release "Queens pheromones" to attract workers. Queens benefit for her care and protection while workers gain information about their queen [20]. There are also recruitment pheromones and royal pheromones. Recruitment pheromones lead to leave the nest and migrate to work. Terrestrial ants use recruitment pheromones to follow the trial. Royal pheromones provide termites to find the queen and attract them to follow the trial [20].

2. Pheromones with plant volatiles

Pheromones and pheromones plus plant volatiles and pests respond to those molecules and is gaining importance [26]. The effectiveness of pheromones can be enhanced by plant volatiles (**Table 1**). In one of the experiments it was mentioned that traps baited with pheromones and host plant, volatiles captured more adults than traps pheromone alone [28]. Aggregation of pheromone ferrugineol with plant volatiles was more effective for the control of *Rhynchophorus ferrugineus* (Coleoptera: Curculionidae) [29]. Green leaf volatiles (GLV) are six carbon atom groups and consist of alcohols, aldehydes, and acetones. They can improve or inhibit the response of insects to their pheromones. For instance, boll weevil aggregation pheromone with GLV-2-hexenol led to more weevils caught in traps in contrast to pheromone alone [34, 35]. Another example of boll weevil traps comprised of aggregation pheromone of *Anthonomus grandis* with green leaf volatiles from cotton plants [30]. Fruit volatile benzaldehyde with plum curculio aggregation pheromone increased the trap captures [31, 36]. Similarly, grandisoic acid which is an aggregation pheromone used for monitoring and control of the plum curculio in apple trees [37]. Deciduous plants volatile with pheromones of corn earworm and the codling moth were also more effective than pheromone alone [27]. But GLVs with pheromones interrupted pheromone responses in the *Tomicus piniperda*, *Conophthorus resinosae, Dendroctonus pseudotsugae,* and *Ips typographus* [34]. (Z)-3-hexenyl acetate is an important chemical to locate the diamondback moth *P. xylostella*'s host [38]. This compound coupled with the synthetic

Insect	Pheromone compound	Host	Plant Volatile	Reference
Helicoverpa zea *Cydia pomonella*	(Z)-3-hexenyl acetate	Corn	Green Leaf Volatile	[27]
Debdroctonus ponderosae	Trans-verbenol Exo-brevicomin	Pine	Myrcene Terpinolene 3-Carene	[28]
Rhynchophorus ferrugineus	Ferrugineol Oryctelure	Palm	Host Palm volatiles	[29]
Anthonomus grandis	Grandlure	Cotton	Six carbon alcohols Aldehydes	[30]
Conotrachelus nenuphar	Grandisoic acid	Apple orchard	Benzaldehyde ethyl isovalerate limonene	[31]
Agrotis ipsilon	Z7-dodecenyl acetate	Linden blossoms	Heptanal	[32]
Helicoverpa armigera	Z11–16:AL Z9–16:AL	Sunflower	Sunflower essential oil dissolved in mineral oil	[33]

Table 1.
Experiments about pheromone and host volatiles.

pheromone resulted the highest response of moths. Antennal receptors show that some olfactory receptor neurons (ORNs) responded to GLVs besides GLV's synergized responses of ORN [39]. Under aphid attack, plants release some compounds which trigger the chemical defense mechanism. These compound attract the aphid predators. In the presence of methyl salicylate the colonization of the aphids was reduced. Methyl salicylate also inhibits males of the *Pieris napi* from mating.

3. Trap design

There are some important properties for trap design. These are height, size, shape, alignment at right angles to the wind, position, and timing of the trap [20]. Some weevils preferred light red color for traps. For efficient usage, traps should be at an exact height such as 50 cm above from the ground [40]. **Figure 1** illustrates four different pheromone baited traps for monitoring Pea Leaf Weevil. Pheromone-baited pitfall and ramp traps caught more adults than delta and ground traps. In both pea and lentils, trap location, type of dispenser, dose, and purity of the pheromone, and environmental conditions will affect the trap catches achievements [41].

In one of the monitoring experiments, the funnel traps were more attractive when compared to adhesive traps for the monitoring of *Palpita unionalis* Hubner (Lepidoptera: Pyralidae) and also more males were caught in the edge instead of the interior part of the forest [42]. Identification, count and removal were easier than adhesive traps and also contamination was less in the funnel trap. It is also mentioned that traps for *Cydia pomonella* located along the border of the orchards were more successful when compared to traps located 30–50 m inside [43]. Location of the trap such as in apple trees traps positioned at 4 m captured more *Cydia pomonella* L.

Figure 1.
Different trap types for Sitona lineatus: (a) Delta Trap, (b) Ground Trap, (c) Ramp Trap, (d) Pitfall Trap [41].

(Lepidopetra: Tortricidae) male compared to those positioned 2 m [44]. Also in one of the experiments traps placed on the top of the tree captured more *Choristoneura rosaceana* Harris (Lepidopetra: Tortricidae) than traps placed in the middle and at the end [45]. One of the experiment groups tested three categories of traps that are bucket style trap, sticky trap, and local trap with three pheromone lure type, they are four-component lure (4C), two-component lure, and three-component lure (3C) for the management of *Spodoptera frugiperda*. In conclusion, bucket-style trap with 3C lures attracted more fall armyworm moths than other combinations with trap and lures. Another experiment concerning trap design for catching *Anoplophora glabripennis* (Coleoptera Cerambycidae) [46]. For monitoring intercept panel traps hung on poplar trees, screen sleeve traps wrapped around poplar trunks, intercept panel traps hung on bamboo poles 20 m away from host trees were used. Male produced pheromone alone and baited with a mixture of (−)-linalool, (Z)-3-hexen-1-ol, linalool oxide, trans-caryophyllene, and trans-pinocarveol. Screen sleeve traps baited with a combination of (−)-linalool and the pheromone caught the highest number of beetles while bamboo poles hung on poplar trees caught the lowest number [47]. It is also mentioned that optimization of pheromone lures and the trap design for monitoring *Dioryctria abietivorella* (Lepidoptera: Pyralidae) is essential for the management of this pest [48]. Three types of sticky trap (White diamond trap, green delta trap, white wing trap) and a green bucket type trap were compared and the most successful trap was found diamond trap while the least males bucket-style bucket style white trap. There are two types of pheromone application methods. One of them is using pheromones directly which includes mass trapping and area-wide dissemination applications. Area-wide dissemination includes mating disruption, attract & kill, and push & pull strategy. The other one is using pheromones indirectly includes monitoring and detecting the spraying time strategy.

4. Insect pheromones with different application techniques

4.1 Population monitoring

This management technique is responsible for maintaining various aspects such as species detection, controlling treatment time, targeting pest dispersion, early warning, and population density and fluctuation, based on the specific needs of the situation. [49, 50]. At the same time, biodiversity hotspots can be identified and species-specific pests can be monitored without risks to non-target organisms with this management strategy [51]. Synthetically derived pheromone was formula into the dispenser and trap to attract the pests. There are different kinds of pheromone dispensers for monitoring insects population such as plastic laminates, twist ties, wax formulations, polyethylene vials, rubber septa, hollow fibers, and impregnated ropes. The benefits of pheromone-based monitoring include affordability, specificity to certain species, and user-friendliness. However, the success of the process also depends on crucial factors such as the rate of pheromone release, the trap's design, its color, and the strategic placement of the traps [42, 48, 52–59]. In the population monitoring technique, to monitor *Conotrachelus nenuphar* (Coleoptera: Curculionidae) apple trees were baited with grandisonic acid (aggregation pheromone) and benzaldehyde (synthetic fruit volatile) to attract curculio adults. Sex pheromones produced by females are used for the monitoring of *Lymantria dispar* L. (Lepidoptera: Lymantriidae) [60], *Heliothis* spp. (Lepidoptera: Noctuidae) populations [61] and Codling moth [62].

Aggregation pheromones usually secreted from males and attract both sexes of pest to mate and feed. They are used to monitor the boll weevil and plum curculio [63–66]. Some of the pheromone compounds, their hosts, and invasive species were given in **Table 2**. But these pheromone compounds should be determined according to the climate conditions, region-specific host plant species, and pest species. The number of males in different traps will give you idea to establish the treatment time. Sex pheromones are useful to determine the exact management time and this will help to reduce the cost of the management strategies. The knowledge about pest and geographical distribution will improve the success of catches. Pheromone traps are useful to monitor for mostly Lepidoptera, Coleoptera, and Diptera Family. Climate changes, moth age, moth behavior, moonlight, pheromone release rate, crop cover, crop phenology effects performance of the monitoring strategy [70–72].

4.2 Mass trapping

The aim of mass trapping tactic is to eliminate males or females with semi chemicals such as host volatiles, aggregation, and sex pheromones from the target pest population to prevent population growth. Trap design, population level, biology of the pest, density, and immigration can change the success of the application. For example, regarding biology of the target pests, mass trapping is more effective when the pest is univoltine, monogamous, and monophagous instead of multivoltine, polygamous, and polyphagous species [73]. The economic viability of mass trapping as a control measure is determined by the composition of pheromones and the expenses incurred in their production. For instance, *Melanotus communis* Gyllenhal (Coleoptera: Elateridae) employs a solitary chemical, 3-tetradecanyl acetate, which influences the cost of traps and the manual labor required, ultimately affecting its competitiveness in the farming industry. In banana crops, the technique of mass trapping was employed to control two types of weevils, such as *Cosmopolitus sardidus* (Coleoptera: Curculionidae) and *Metamosius hemipterus* (Coleoptera: Curculionidae), using distinct methods. The former was captured using a pitfall trap baited with a

Insect	Pheromone compounds	Host	Reference
Dioryctria abietivorella	(9Z,11E)-tetradecadien-1-yl acetate (3Z,6Z,9Z,12Z,15Z)-pentacosapentaene	Breeding orchards	[48]
Acleris variona *Acleris gloverana*	E-11-13-14:Ald	Pinaceae	[49]
Synanthedon scitula	(Z,Z)-3,13-octadecadien-1-ol acetate (ODDA)	Apple orchards	[52]
Choristoneura fumiferana	(E:Z)-11-tetradecenal	Picea plauca	[59]
Scrobipalpa atriplicella	(Z)-5 dodecenyl acetate €-5-dodecenyl acetate	Scrobipalpa atriplicella	[67]
Contarinia nasturtii	(2S,9S)-diacetoxyundecane (2S,10S)diacetoxyundecane (2S)-acetoxyundecane	Crucifers	[68]
Heliothis armigera	Z-11-Hexadecenal Z-9-Hexadecenal	Semi-arid tropic crops	[69]

Table 2.
Monitoring trials against invasive pests.

pheromone, while the latter was caught with a gallon trap baited with sugarcane and a pheromone. In plots, corn damage diminished by 61–64% during the experiment. Banana bunch weight increased 23% compared to control plots after 12 months of trapping. Trapping for *C. sordidus* over 200 hectares reduced corn damage up to 62–86%. Insecticide control measures resulted in about 20–30% corn damage while the use of pheromone trapping to manage *C. sardidus* lowered corn damage to 10% [74]. Optimization of the lures and traps [73], operational costs provide to determine the most suitable method [75, 76] but, inadequate number of traps, polygamous nature of codling moths leads to inadequate control of pests [73]. Surface lubricants [77, 78], surface conditioners (Fluon), insecticides, water, or other physical structures, increase the number of pests collected in traps [18, 78]. Lures that attract both male and female [79] by using one trap to collect different pests will improve the success of mass trapping [29].

Chinese first mass trapping test using rubber septa containing pheromones of *Grapholita molesta* was found successful. They observed that female mating rates were diminished 74% and 83% in treated orchards compared to control and fruit infestation rates per fruit reduced by 50–70% and thus reduce 50% reduction in cost of insecticide application. In China, a research group found that pheromone compounds identified for *Paranthrene tabaniformis* in the United States was not effective in the field of China. It was due to the Chinese strain of pheromones consist of compounds in different ratios when compared to United States strain [80]. The field trapping experiments were carried out by various types of trapping systems and different dosages of pheromone lures. At the end, mass trapping system resulted in a significant reduction of this pest with female mating rates decreased by 43–79% and the total pest population reduced by around 57–95%. The use of mass trapping technique was experimented on the overwintering generation of C. suppressalis on rice plants. The results showed a significant decrease of 61% in egg laying and a 57% decrease in rice plant infestation in the treated field, as compared to the field treated with pesticides. [81]. Experiments about some of the pests their host and region was given in **Table 3**.

Aggregation pheromones of predators can be used for mass trapping such as aggregation pheromones of *Coccinella septempunctata* L. (Coleoptera) can be used for the management of aphids [84]. Recently this mass trapping tactic is successful for the management of the Japanese beetle, *Popillia japonica* Newman (Coleoptera: Scarabaidae) [85].

Insect	Order	Host	Region	Reference
Cosmopolites sordidus *Metamasius hemipterus*	Coleoptera: Curculionidae	Banana	Costa Rica	[74]
Synanthedon scitula	Lepidoptera: Sesidae	Apple Orchards	Nort Carolina Virginia	[75]
Rhynchophorus ferrugineus	Coleoptera: Curculionidae	Palm	Middle East	[29]
Coryphodema tristis	Lepidoptera: Cossidae	*Eucalyptus nitens*	South Africa	[82]
Ectropis grisescens	Lepidoptera: Geometridae	Tea	China	[83]

Table 3.
Mass trapping trials against invasive pests.

4.3 Attract and kill

Attract and kill method used for capturing and killing the pests. Both field crop and stored product pests are controlling with this method [20]. The attractive method shares similarities with mass trapping, but its effectiveness extends to a broader area and its killing impact is not limited to a single trap [86]. When pest density and immigration rate is low, attract and kill method is working well similar to mass trapping method [73, 87, 88]. For some of the Lepidoptrean pests, attract and kill method is better when compared to mating disruption because attract and kill method does not allow males to recover or mate [89]. In the management strategy *Carpophilus* spp. aggregation pheromone was incorporated with ripening fruit and insecticide [90]. *A. grandis* that is a main pest of cotton in the USA and males produce aggregation pheromone grandlure [91]. This aggregation pheromone and control tubes provided strong reduction in pest population [66, 92]. *Musca domestica* are attracted by muscalure used with co-attractants in attract and kill approaches [93, 94]. In Switzerland 0.16% pheromone and 6% permethrin were used to control codling moth as part of the attract and kill method. Attract and kill method have been effective for *Bactrocera dorsalis*, *Bactrocera cucurbitacea* and *C. capitata* for 50 years [95]. Furthermore there are lots of studies about this attract and kill method (**Table 4**).

4.4 Mating disruption

Mating disruption prevents pests to find their mates and hinders their reproduction of them [89, 101–104]. There are four mechanisms that should take into consideration, one of them is semi chemicals leads to false trail by attracting males away from the females while the other mechanism is camouflage. The presence of semi chemicals pervades the environment, hindering the ability of males to locate females and thereby disrupting the mating process between male and female. Sensory desensitization by overexpressed semi chemicals also leads to hinder males to find females. The last mechanism leads to the emigration of males due to the excess pheromone release. Male cannot find females for mating [20, 105, 106]. Application rate, dispenser design, dispenser height, and trap density are important components for the success of the management technique. There are some important factors that should be considered due to several mating disruption dispensers and formulations. Passive dispensers are leading the way in mating disruption programs due to

Insect	Order	Host	Region	Reference
Pectinophora gossypiella	Lepidoptera: Gelechiidae	Cotton	California Arizona	[96, 97]
Spodoptera littoralis	Lepidoptera: Noctuidae	Cotton	Egypt	[98]
Cydia pomonella	Lepidoptera: Tortricidae	Apple	Switzerland	[88]
Ephestia kuehniella	Lepidoptera: Pyralidae	Stored Product	Italy	[99]
Ectropis grisescens	Lepidoptera: Geometridae	Tea	China	[83]
Carpophilus spp.	Coleoptera: Nitidulidae	Stone fruit orchards	Australia	[100]

Table 4.
Attract and kill trials against invasive pests.

their ease of application and cost-effectiveness compared to other options available. But they release pheromones continuously. Time of the day and pest flight activity time should be considered. Aerosol dispensers can release pheromones at the exact time, and the amount of pheromones released per unit can be arranged. Aerosol delivery system is faster and cheaper than passive dispensers and never be affected by environmental degradation. Aerosol dispensers target multiple pest species and modern digital electronic technologies support this technique. But wind can affect the efficiency of aerosols and this technique is more effective when applied for large areas [104]. In 1953 China, *C. pomonella* pheromone lures were placed in apple orchards. After 2 years, the mating rates of female moths reduced by 55–75% and the infestation rate also decreased by 35–77%. This was the first study using mating disruption in China [107, 108].

This technique is environmentally benign, does not affect non-target organisms, and is approved for organic production system [18]. Mating disruption is a successful management strategy for *L. dispar* [109], *Cydia pomonella* [110], *Lobesia botrana* [111], *Grapholita molesta* [112], *Ephestia cautella*, *Ephestia kuehniella*, *Plodia interpunctella* [113], *Cossus insularis* [114], *Grapholita molesta*, *Keiferia lycopersicella*, *Pectinophora gossypiella*, *Platynota stultana* [11, 35, 115], *Tuta absoluta* [116], *Plodia interpunctella* (Wjayaratne), *Chilo suppressalis* [117], *Cydia fagiglandana* and *C. splendana* [118]. There are also non-lepidopteran pests *Anomala orientalis, Planococcus ficus* [119], and *Prionus californicus* that are affected by mating disruption [120, 121] (**Table 5**).

Population density is important for the success of this management technique. High density of the population is difficult to control [102, 124–127]. In one of the studies Borchert and Walgenbach (2000) mentioned that mating disruption was effective in low-density populations of the *Platynota idaeusalis* [128]. It is also reported that mating disruption application with pesticides or alone did not control *Cydia pomonella* in high-density population [126]. Attractive plant-derived kairomones with sex-releasing pheromones increased the effectiveness of mating disruption technique [129–132]. Other successful mating disruption trials in low density have been experimented for *Lymantria dispar* [35, 133, 134] and *Cydia splendana* [35, 135]. Especially landscapes such as residential areas, parks, commercial sites, with low density of pest, mating disruption technique are effective. Mating disruption technique was not found effective when males and females were close to each other in space and time [133]. The amount of pheromone lures is also important to increase moth captures. For example, it is observed that the trapping of codling moth males

Insect	Order	Host	Region	Reference
Tuta absoluta	Lepidoptera: Gelechiidae	Tomato	Turkey	[116]
Chilo suppressalis	Lepidoptera: Pyralidae	Rice	China	[117]
Cydia fagiglandana *Cydia splendana*	Lepidopetra: Tortricidae	Chestnut	Italy	[118]
Planococcus ficus	Hemiptera: Pseudococcidae	Grape	Italy	[119]
Plodia interpunctella	Lepidoptera: Pyralidae	Stored Product	USA	[122]
Lobesia botrana	Lepidoptera: Tortricidae	Grape	Spain	[123]

Table 5.
Mating disruption trials against invasive pests.

increased when 10–20 mg pheromone was baited [9] or by position traps in the upper canopy [11, 35]. Exposure to a high concentration of pheromone blend lead to generating insensitive males [110] and immigration have an adverse impact on the success of mating disruption [102].

Numerous methods are utilized in mating disruption to dispense pheromones. Examples include microencapsulation, hand-applied dispensers, polymer spirals, and ropes, spray application similar to insecticides, as well as hollow fibers and twist-tie ropes. New devices which have large polymer bags loaded with large amounts of pheromones are used instead of batteries and other puffers. Pheromone release rate, different dispenser types, dispenser placement, or population density will affect the control of pests. Nevertheless, mating disruption is an excellent technique for integrated pest management programs.

In another experiment, mating disruption technique was used to control *Tuta absoluta*. To assess the efficacy of pheromone lures, six fields were monitored, with three fields treated with pheromones and three fields left as control. The number of male insects detected in traps per week was significantly lower in pheromone-treated fields compared to the control groups in both 2018 and 2019. Specifically, the mean number of males detected per week was 120 ± 16 and 69 ± 15 in the pheromone-treated fields, while it was 299 ± 16 and 230 ± 15 in the control groups, respectively. Moreover, the percentage of infestation rate was significantly lower in pheromone-treated fields than in control fields in both years. Specifically, the mean infestation rate was 4 ± 0.56% and 1 ± 0.52% in the pheromone-treated fields, while it was 11 ± 0.56% and 7 ± 0.52% in the control fields in 2018 and 2019, respectively [116].

In Southeast Germany, Ecoflex fibers have been experimented with in mating disruption trials for the management of *Lobesia botrana* [136]. Later electrospun mesofibers were became popular with their biodegradable property and also harmless to non-target organisms [137]. Nowadays, Isonet L TT was tried for 3 years in Italy and found to be successful for the control of the *L. botrana* during the whole season [138].

4.5 Push-pull strategies

Push-pull strategy is first coined by Pyke and coworkers. Repellent and stimuli were used for the management of *Helicoverpa* species in cotton [139]. Currently, this tactic is also called a "stimulo-deterrent diversion tactic." There are two types of push-pull strategies, one of them is intercropping with repulsive non-crop plants with attractive trap crop and the other one is semi chemical repellents with attractive pheromone traps. Repellents cause pests to avoid the crop (push) and attracting pests to pheromone traps that remove pests before they find mates or hosts (pull). Insect biology, chemical ecology, and interaction between host plants and natural enemies should be known to apply this strategy [140–145]. Synthetic repellents, host and non-host volatiles, host-derived semi chemicals, antifeedants, oviposition deterrents, and stimulants can be used in these strategies. Traps capture rate depends on (1) trap size and efficiency, (2) traps blends of attractive volatiles, (3) release rates, (4) population density of flying pest insects, and (5) competitive sources of natural attraction. There are some experiments about push and pull strategies for the management of the Lepidoptera family (*Helicoverpa armigera*, *S. frugiperda*), Coleoptera family (*Sitona lineatus, Leptinotarsa decemlineata, Meligethes aeneus, Dendroctonus ponderosae*), and Diptera family (*Drosophila suzukii*) which were given in **Table 6**.

Insect	Order/Family	Host	Region	Reference
Helicoverpa armigera	Lepidoptera: Noctuidae	Cotton	Australia	[121]
Sitona lineatus	Coleoptera: Curculionidae	Bean	England	[146]
Leptinotarsa decemlineata	Coleptera: Chrysomalidae	Potato	Maine	[147]
Meligethes aeneus	Coleoptera: Nitidulidae	Oilseed Rape	UK	[148]
Dendroctonus ponderosae	Coleoptera: Curculionidae	Pine	British Colombia	[28]
Drosophila suzukii	Diptera: Drosophilidae	Raspberry	USA	[149]
Spodoptera frugiperda	Lepidoptera: Noctuidae	Maize	Uganda, Tanzania, Kenya	[150]

Table 6.
Push-Pull trials against invasive pests.

The behavior of pests is manipulated by push-pull tactics by combining stimuli for push components and stimuli for pull components. To deter pests, various methods can be employed such as altering the host's color, shape, or size. DEET (N, N-diethyl-meta-toluamide) is commonly used for warding off cockroaches and lady beetles. On the other hand, volatiles like citronella and eucalyptus essential oils can be utilized to mask host odors or induce non-host avoidance. Semi chemicals, antifeedants such as azadirachtin, alarm pheromones (generally used for aphids) are also used for push components. Visual stimulants, host volatiles (Baits, HIPVs), kairomones-post volatiles, sex and aggregation pheromones, and also gustatory and oviposition stimulants can be used as stimuli for pull components. In one of the case studies to protect the maize and sorghum crops from stem borer pests. Stem borers are repelled from the crops by planting repellent nonhost intercrops like molasses grass (*Melinis minutiflora*), silverleaf desmodium (*Desmodium uncinatum*), or greenleaf desmodium (*Desmodium intortum*) which set as the "push." Then trap crops like Napier grass (*Pennisetum purpureum*) or Sudan grass (*Sorghum vulgare sudanense*)—the "pull" component are used. For the management of *Helicoverpa* in cotton, neem seed extracts are applied to the cotton crop as push strategy and pigeon pea (*Cacanus cajan*) or maize (*Zea mays*) is planted as a pull strategy. A pull strategy was employed for managing the Colorado potato beetle, wherein the beetle was attracted to host plant volatiles and a potato trap crop was sprayed with an attractant. Additionally, a push strategy was implemented by applying antifeedant neem between the rows.. Then instead of plant volatile Colorado beetle aggregation pheromone (S)-3,7-dimethyl-2-oxo-6-octene-1,3-diol was applied as pull strategy. *S. lineatus* pea leaf weevil is a pest of legumes its pheromone 4-methyl-3,5-heptanedione used as pull and commercially available neem antifeedant used as push component of the strategy. Another case study was about the management of pollen beetle *M. aeneus* which is acted on oilseed rape (*Brassica napus*). Turnip rape was planted as the pull stimuli and lavender (*Lavandula angustifolia*) was used as a push component and also alkenyl glucosinolates were used as the main attractant of such pest. Another example for a pull and push strategy is *Delia antiqua*. It is a main pest of onion. Small unmarketable bulbs were used as a trap crop as a pull strategy, cinnamaldehyde is an oviposition deterrent used as a push stimulus. So significant reduction was observed when using this push pull strategy. For the management of *Drosophila suzukii*, mass trapping was applied as a pull strategy and

1-octen-3-ol which is an oviposition deterrent was used as a push strategy. In the end high reduction in oviposition of *D. Suzukii* was observed.

There are lots of advantages of push-pull tactics over traditional pest management methods. This strategy can be used both in juveniles and adult stages of pests. Simple, commercially available, nontoxic, and cheap components can be used as a stimuli. This management technique can increase efficiency of individual push and pull components, antifeedants and oviposition deterrents are used as a push tactic and it do not need to struggle with resistance management. Besides there are some disadvantages such as limited specificity, and there are lots of odor sources for attraction. We should develop the technique by understanding the behavioral and chemical ecology of the host pest, and the development of semi chemical components. People generally choose to use insecticide applications instead of biological control agents. But this strategy is a useful tool for integrated pest management programs reducing pesticide input [151]. Enhancement of monitoring and decision-making systems is crucial, and a comprehensive approach is required to regulate the system [152].

The presence of Nepetalactone, an aphid sex pheromone constituent, and (Z)-Jasmone, a plant volatile, can have an impact on aphid parasitoids. In turn, these parasitoids can offer a natural means of controlling aphid populations. While tricosane and pentacosane lady beetle pheromones provide to push the parasitoids from surrounding areas to the treatment crop. These two lady beetle pheromone components (tricosane and pentacosane) used by the *Aphidius ervi* which is an aphid parasitoid to escape from seven spotted lady beetle [153].

Another example for push-pull strategy is about bark beetles (Scolytidae). Antiaggregation pheromone 3-methylcyclohex-2-en-1-one and aggregation pheromone (frontalin, seudenol, 1-methylcyclohex-2-enol and ethanol) diminished populations of the *D. pseudotsugae* [154]. Push-pull strategy is also applied with aggregation and anti-aggregation pheromones for the management of other forest pests that are *D. ponderosae, Ips paraconfusus*, *Dendroctonus frontalis* [142]. "Verbenone" is an anti-aggregation pheromone that reduces the attract rates of *D. ponderosae* Hopkins (Coleoptera: Curculionidae) [155–159]. Trees baited with aggregation pheromone grandisoic acid and fruit volatile benzaldehyde attracted plum curculio adults. It is also mentioned that instead of stand art insecticide application pheromone-baited "trap" trees provided satisfactory suppression of fruit injury [37]. As a result, 93% fewer trees were sprayed with insecticide with this "trap" tree approach [37].

Chemical ecology, geographical variation can change aggregation pheromone components [160]. For instance Australian and the Hawaiian populations of the *Rhabdoscelus obscurus* Boisduval produce male specific 2-methyl-4-octanol. This compound can enhance the attractiveness of Hawaiian populations only. Australian population (E2)-6-methyl-2-hepten-4-ol (rhynchophorol) combined with 2-methyl-4-octanol to attract the Australian males. This experiment shows that interactions between organisms and their pheromone according to their regions should be studied.

5. Conclusion

The intensive use of pesticides leads to residue problems on food, resistance of pests, environmental pollution, and health problems. Pheromones can be used safely instead of broad-spectrum insecticides in sustainable management strategies. Pheromones are volatile and do not leave residue behind, species-specific and cost-effective and do not cause toxicity. By using pheromones, male adult number, mating rates can be reduced and also insecticide application time can be detected by monitoring strategy. Nowadays studies are mainly about identification, optimization of pheromone components of region-specific species, and also development of new formulations. Active compounds extracted from insect frass can be used as repellent or deterrent for the conspecifics. Besides new strategy is about odorant receptor genes which allows discovery of compounds that can trigger or block these receptors. Plant-insect interactions also should be studied and enzymes involved in the odorants and pheromones degradation should be observed to interfere the chemical communication of insect. The ultimate change will be to improve the process by employing simulation models. One factor of concern is high cost of the pheromones in comparison to pesticide application. For this reason producers, farmers and scientists should reduce the cost of application of pheromones to affordable threshold. Hence they should target low population density, to release pheromone in exact time of flight, to arrange the pheromone release rate, to develop more efficient formulations regarding the ecological conditions and also they should target to use multiple pheromones to monitor several pest simultaneously. Successful implementation of pheromone based management techniques can diminish economic cost and lighten the ecological footprints as this management technique decreases insecticide usage.

Conflict of interest

The authors declare no conflict of interest.

Author details

Melis Yalçın
Faculty of Agriculture Engineering, Adnan Menderes University, Aydın, Turkey

*Address all correspondence to: melisusluy@adu.edu.tr

References

[1] Beth A. Neglected hormone. Natural Science. 1932;**20**:177-183

[2] Karlson P, Butenandt A. Pheromones (ectohormones) in insects. Annual Review of Entomology. 1959;**4**:39-58. DOI: 10.1146/annurev.en.04.010159.000351

[3] Karlson P, Lüscher M. "Pheromones" a new term for a class of biologically active substances. Nature (London). 1959;**183**:55-56. DOI: 10.1038/183055a0

[4] Butenandt VA. Uber den sexsual-lockstoff des seidenspinners Bombyx mori. Reindarstellung und Konstitution. Zeitschrift für Naturforschung. Teil B. 1959;**14**:283

[5] Petkevicius K, Löfstedt C, Borodina I. Insect sex pheromone production in yeasts and plants. Current Opinion in Biotechnology. 2020;**65**:259-267. DOI: 10.1016/j.copbio.2020.07.011

[6] Rizvi SAH, George J, Reddy GVP, Zeng X, Guerrero A. Latest developments in insect sex pheromone research and its application in agricultural pest management. Insects. 2021;**12**(6):484. DOI: 10.3390/insects12060484

[7] El-Sayed AM, Ganji S, Gross J, Giesen N, Rid M, Lo PL, et al. Climate change risk to pheromone application in pest management. The Science of Nature. 2021;**108**:47. DOI: 10.1007/s00114-021-01757-7

[8] Turgut C. The contamination with organochlorine pesticides and heavy metals in surface water in Küçük Menderes River in Turkey, 2000-2002. Environment International. 2003;**29**(1):29-32. DOI: 10.1016/S0160-4120(02)00127-7 PMID: 12605933

[9] Gut LJ, Stelinski LL, Thomson DR, Miller JR. Behaviour-modifying chemicals: Prospects and constraints in IPM. In: Koul O, Dhaliwal GS, Cuperus GW, editors. Integrated Pest Management: Potential Constraints and Challenges. New York, NY: CABI; 2004. pp. 73-121. DOI: 10.1079/9780851996868.0073

[10] Millar JG, McElfresh JS, Romero C, Vila M, Marí-Mena N, Lopez-Vaamonde C. Identification of the sex pheromone of a protected species, the Spanish moon moth Graellsia isabellae. Journal of Chemical Ecology. 2010;**36**:923-932. DOI: 10.1007/s10886-010-9831-1

[11] Welter SC, Pickel C, Millar J, Cave F, Steenwyk RAV, Dunley J. Pheromone mating disruption offers selective management options for key pests. California Agriculture. 2005;**59**(1):16-22. DOI: 10.3733/ca.v059n01p16

[12] Ando T, Inomata SI, Yamamoto M. Lepidopteran sex pheromones. In: Schulz S, editor. The Chemistry of Pheromones and Other Semiochemicals I, Topics in Current Chemistry. Vol. 239. Berlin/Heidelberg, Germany: Springer; 2004. pp. 51-96. DOI: 10.1007/b95449

[13] Löfstedt C, Wahlberg N, Millar JM. Evolutionary patterns of pheromone diversity in Lepidoptera. In: Allison JD, Cardé RT, editors. Pheromone Communication in Moths: Evolution, Behavior and Application. Berkeley, CA, USA: University of California Press; 2016. pp. 43-78. DOI: 10.1525/9780520964433-005

[14] Yew JY, Chung H. Insect pheromones: An overview of function, form, and discovery. Progress in Lipid

Research. 2015;**59**:88-105. DOI: 10.1016/j.plipres.2015.06.001

[15] Naka H, Fujii T. Chemical divergences in the sex pheromone communication systems in moths. In: Ishikawa Y, editor. Insect Sex Pheromone Research and Beyond. Singapore: Springer; 2020. pp. 3-17. DOI: 10.1007/978-981-15-3082-1_1

[16] Ginzel MD. Olfactory signals. In: Breed M, Moore J, editors. Encyclopedia of Animal Behavior. Vol. 2. Oxford, UK: Elsevier Ltd.; 2010. pp. 584-588

[17] Law J, Regnier F. Pheromones. Annual Review of Biochemistry. 1971;**40**:533-548. DOI: 10.1146/annurev.bi.40.070171.002533

[18] Tewari S, Leskey TC, Nielsen AL, Pinero JC, Rodriguez-Saona CR. Use of pheromones in insect pest management, with special attention to weevil pheromones. In: Abrol DP, editor. Integrated Pest Management. New York, USA: Elseiver; 2007. pp. 141-168. DOI: 10.1016/B978-0-12-398529-3.00010-5

[19] Vandermoten S, Mescher MC, Francis F, Haubruge E, Verheggen FJ. Aphid alarm pheromone: An overview of current knowledge on biosynthesis and functions. Insect Biochemistry and Molecular Biology. 2012;**42**(3):155-163. DOI: 10.1016/ j.ibmb.2011.11.008Ghany

[20] Abd El-Ghany NM. Semiochemicals for controlling insect pests. Journal of Plant Protection Research. 2019;**59**(1):1-11. DOI: 10.24425/jppr.2019.126036

[21] Regnier FE, Law JH. Insect pheromones. Journal of Lipid Research. 1968;**9**:541-551. DOI: 10.1016/S0022-2275(20)42699-9

[22] Weatherston I, Minks AK. Regulation of semiochemicals – global aspects. Integrated Pest Management Reviews. 1995;**1**(1):1-13. DOI: 10.1007/BF00140330

[23] Tinzaara W, Dicke M, van Huis A, Gold CS. Use of infochemicals in pest management with special reference to the banana weevil, Cosmopolites sordidus (Germar) (Coleoptera: Curculionidae). Insect Science and Its Application. 2002;**22**:241-261. DOI: 10.1017/S1742758400020877

[24] Blomquist GJ, Figueroa-Teran R, Aw M, Song M, Gorzalski A, Abbott NL, et al. Pheromone production in bark beetles. Insect Biochemistry and Molecular Biology. 2010;**40**:699-712. DOI: 10.1016/j.ibmb.2010.07.013

[25] Prokopy R, Ziegler J, Wong T. Deterrence of repeated oviposition by fruit-marking pheromone in Ceratitis capitata (Diptera: Tephritidae). Journal of Chemical Ecology. 1978;**4**:55-63. DOI: 10.1007/BF00988260

[26] Baker TC, Heath JJ. Pheromones—function and use in insect control. In Gilbert L I. Iatro K. Gill SS, editors. Molecular Insect Science. Amsterdam, The Netherlands. Elsevier; 2004. p. 407-460. DOI:

[27] Light DM, Flath RA, Buttery RG, Zalom FG, Rice RE, Jung EB. Host plant green leaf volatiles synergize the synthetic sex pheromones of the corn ear worm and codling moth (Lepidoptera). Chemoecology. 1993;**4**:145-152. DOI: 10.1007/BF01256549

[28] Borden JH, Pureswaran DS, Lafontaine JP. Synergistic blends of monoterpenes for aggregation pheromones of the mountain pine beetle (Coleoptera: Curculionidae). Journal of Economic Entomology. 2008;**101**:1266-1275. DOI: 10.1093/jee/101.4.1266

[29] Hallett RH, Oehlschlager AC, Borden JH. Pheromone trapping protocols for the Asian palm weevil, Rhynchophorus ferrugineus (Coleoptera: Curculionidae). International Journal of Pest Management. 1999;**45**:231-237. DOI: 10.1080/096708799227842

[30] Dickens J. Green leaf volatiles enhance aggregation pheromone of boll weevil, *Anthonomus grandis*. Entomologia Experimentalis et Applicata. 1989;**52**:191-203. DOI: 10.1111/j.1570-7458.1989.tb01268.x

[31] Piñero JC, Prokopy RJ. Field evaluation of plant odor and pheromonal combinations for attracting plum curculios. Journal of Chemical Ecology. 2003;**29**:2735-2748. DOI: 10.1023/B:JOEC.0000008017.16911.aa

[32] Hoffmann A, Bourgeois T, Munoz A, Anton S, Gevar J, Dacher M, et al. A plant volatile alters the perception of sex pheromone blend ratios in a moth. Journal of Comparative Physiology. 2020;**206**:553-570. DOI: 10.1007/s00359-020-01420-y

[33] Ian E, Kirkerud NH, Galizia CG, Berg BG. Coincidience of pheromone and plant odor leads to sensory plasticity in the heliothine olfactory system. PLoS One. 2017, 2017;**12**(5):e0175513

[34] Aldrich JR, Bartelt RJ, Dickens JC, Knight AL, Light DM, Tumlinson JH. Insect chemical ecology research in the United States Department of Agriculture–Agricultural Research Service. Pest Management Science. 2003;**59**:777-787. DOI: 10.1002/ps.710

[35] Reddy GVP, Guerrero A. New pheromones and inset control strategies. In: Vitamin and Hormones. Vol. 83. Elseiver; 2010. pp. 493-519. DOI: 10.1016/S0083-6729(10)83020-1

[36] Piñero JC, Wright SE, Prokopy RJ. Response of plum curculio (Coleoptera: Curculionidae) to odor-baited traps near woods. Journal of Economic Entomology. 2001;**94**:1386-1397. DOI: 10.1603/0022-0493-94.6.1386

[37] Leskey TC, Pinero JC, Prokopy RJ. Odor-baited trap trees: A novel management tool for plum curculio (Coleoptera: Curculionidae). Journal of Economic Entomology. 2008;**101**:1302-1309. DOI: 10.1603/0022-0493(2008)101[1302:ottanm]2.0.co;2

[38] Reddy GVP, Guerrero A. Optimum timing of insecticide applications against diamondback moth Plutella xylostella in cole crops using threshold catches in sex pheromone traps. Pest Management Science. 2001;**57**:90-94. DOI: 10.1002/1526-4998(200101)57:1<90::AID-PS258>3.0.CO;2-N

[39] Ochieng SA, Park KC, Baker TC. Host plant volatiles synergize responses of sex pheromone-specific olfactory receptor neurons in male Helicoverpa zea. Journal of Comparative Physiology. A, Neuroethology, Sensory, Neural and Behavioral Physiology. 2002;**188**:325-333. DOI: 10.1007/s00359-002-0308-8

[40] Reddy GVP, Gadi N, Taianao AJ. Efficient sex pheromone trapping: Catching the sweetpotato weeviş, Cylas formicarius. Journal of Chemical Ecology. 2012;**38**:846-853. DOI: 10.1007/s10886-012-0160-4

[41] Reddy GVP, Shrestha G, Miller DA, Oehlschlager AC. Pheromone-trap monitoring system for pea leaf weevil, *Sitona lineatus*: Effects of trap type, lure type and trap placement within fields. Insects. 2018;**9**(3):75. DOI: 10.3390/insects9030075

[42] Athanassiou C, Kavallieratos N, Mazomenos B. Effect of trap type, trap

color, trapping location, and pheromone dispenser on captures of male *Palpita unionalis* (Lepidoptera: Pyralidae). Journal of Economic Entomology. 2004;**97**:321-329. DOI: 10.1093/jee/97.2.321

[43] Knight AL. Influence of within-orchard trap placement on catch of codling moth (Lepidoptera: Tortricidae) in sex pheromone–treated orchards. Environmental Entomology. 2007;**36**(2):425-432. DOI: 10.1093/ee/36.2.425

[44] Epstein DL, Stelinski LL, Miller JR, Grieshop MJ, Gut J. Effects of reservoir dispenser height on efficacy of mating disruption of codling moth (Lepidoptera: Tortricidae) in apple. Pest Management Science. 2011;**67**:975-979. DOI: 10.1002/ps.2141

[45] Agnello AM, Reissig WH, Spangler SM, Charlton RE, Kain DP. Trap response and fruit damage by obliquebanded leafroller (Lepidoptera: Tortricidae) in pheromone-treated apple orchards in New York. Environmental Entomology. 1996;**25**:268-282. DOI: 10.1093/ee/25.2.268

[46] Robert L, Jr M, Agboka K, Tounou AK, Koffi D, Agbevohia KA, et al. Comparison of oheromone trap design and lures for *Spodoptera frugiperda* in Togo and genetic characterization of moths caught. Entomologia Experimentalis et Applicata. 2019;**167**:507-516. DOI: 10.1111/eea.12795

[47] Nehme ME, Keena MA, Zhang A, Baker TC, Xu Z, Hoover K. Evaluating the use of male-produced pheromone components and plant volatiles in two trap design to monitor *Anoplophora glabripennis*. Environmental Entomology. 2010;**39**(1):169-176. DOI: 10.1603/EN09177

[48] Strong WB, Millar JG, Grant GG, Moreira JA, Chong JM, Rudolph C. Optimization of pheromone lure and trap design for monitoring the fir coneworm, Dioryctria abietivorella. Entomologia Experimentalis et Applicata. 2008;**126**:67-77. DOI: 10.1111/j.1570-7458.2007.00635.x

[49] Nealis VG, Silk P, Turnquist R, Wu J. Baited pheromone traps track changes in populations of western blackheaded budworm (Lepidoptera: Tortricidae). The Canadian Entomologist. 2010;**142**:458-465. DOI: 10.4039/n10-033

[50] Wall C. Principle of monitoring. In: Ridgway RL, Silverstein RM, Inscoe MN, editors. Behavior Modifying Chemicals for Insect Management. New York, NY: Marcel Dekker, Inc.; 1990. pp. 9-23. DOI: 10.33338/ef.83484

[51] Larsson MC. Pheromones and other semiochemicals for monitoring rare and endangered species. Journal of Chemical Ecology. 2016;**42**:853-868. DOI: 10.1007/s10886-016-0753-4

[52] Bergh JC, Leskey TC, Sousa JM, Zhang A. Diel periodicity of emergence and premating reproductive behaviors of adult dogwood borer (Lepidoptera: Sesiidae). Environmental Entomology. 2006;**35**:435-442. DOI: 10.1603/0046-225X-35.2.435

[53] Diaz-Gomez O, Malo EA, Patino-Arrellano SA, Rojas JC. Pheromone trap for monitoring *Copitarsia decolora* (Lepidoptera: Noctuidae) activity in cruciferous crops in Mexico. Florida Entomologist. 2012;**95**:602-609. DOI: 10.2307/23268483

[54] Hoddle MS, Millar JG, Hoddle CD, Zou Y, JS ME, Lesch SM. Field optimization of the sex pheromone of *Stenoma catenifer* (Lepidoptera: Elachistidae): Evaluation of lure types, trap height, male flight distances, and number of traps needed per avocado

orchard for detection. Bulletin of Entomological Research. 2011;**101**:145-152. DOI: 10.1017/S0007485310000301

[55] Isaacs R, Van Timmeren S. Monitoring and temperature-based prediction of the whitemarked tussock moth (Lepidoptera: Lymantriidae) in blueberry. Journal of Economic Entomology. 2009;**102**:637-645. DOI: 10.1603/029.102.0223

[56] Knight AL, Fisher J. Increased catch of codling moth (Lepidoptera: Tortricidae) in semiochemical-baited orange plastic delta-shaped traps. Environmental Entomology. 2006;**35**:1597-1602. DOI: 10.1093/ee/35.6.1597

[57] Leskey TC, Wright SE, Short BD, Khrimian A. Development of behaviorally-based monitoring tools for the brown marmorated stink bug (Heteroptera: Pentatomidae) in commercial tree fruit orchards. Journal of Entomological Science. 2012;**47**:76-85. DOI: 10.18474/0749-8004-47.1.76

[58] Roubos CR, Liburd OE. Effect of trap color on captures of grape root borer (Lepidoptera: Sesiidae) males and non-target insects. Journal of Agricultural Urban Entomology. 2008;**25**:99-109. DOI: 10.3954/1523-5475-25.2.99

[59] Sanders C. Evaluation of high-capacity, nonsaturating sex-pheromone traps for monitoring population densities of spruce budworm (Lepidoptera: Tortricidae). The Canadian Entomologist. 1986;**118**:611-619. DOI: 10.4039/Ent118611-7

[60] Kolodny-Hirsch DM, Schwalbe CP. Use of disparlure in the management of the gypsy moth. In: Ridgway RL, Silverstein M, Inscoe MN, editors. Behavior-Modifying Chemicals For Insect Management. New York, USA: Marcel Dekker, Inc.; 1990. pp. 363-385

[61] Lopez JD, Shaver TN, Dickerson WA. Population monitoring of Heliothis. spp. using pheromones. In: Ridgway RL, Silverstein RM, Inscoe MN, editors. Behavior-Modifying Chemicals For Insect Management. New York, NY: Marcel Dekker, Inc.; 1990. pp. 473-496. DOI: 10.33338/ef.83484

[62] Knight AL, Light DM. Developing action thresholds for codling moth (Lepidoptera: Tortricidae) with pear ester and codlemone baited traps in apple orchards treated with sex pheromone mating disruption. The Canadian Entomologist. 2005;**137**:739-747. DOI: 10.4039/n05-040

[63] Kroschel J, Zegarra O. Attract-and-kill: A new strategy for the management of the potato tuber moths *Phthorimaea operculella* (Zeller) and *Symmetrischema tangolias* (Gyen) in potato: Laboratory experiments towards optimising pheromone and insecticide concentration. Pest Management Science. 2010;**66**:490-496. DOI: 10.1002/ps.3483

[64] Leskey TC, Wright SE. Monitoring plum curculio, *Conotrachelus nenuphar* (Coleoptera: Curculionidae), populations in apple and peach orchards in the mid-Atlantic. Journal of Economic Entomology. 2004;**97**:79-88. DOI: 10.1603/0022-0493-97.1.79

[65] Pinero JC, Agnello AM, Tuttle A, Leskey TC, Faubert H, Koehler G, et al. Effectiveness of odor-baited trap trees for plum curculio (Coleoptera: Curculionidae) monitoring in commercial apple orchards in the northeast. Journal of Economic Entomology. 2011;**104**:1613-1621. DOI: 10.1603/EC10310

[66] Ridgway RL, Inscoe MN, Dickerson WL. Role of the boll weevil

pheromone in pest management. In: Ridgway RL, Silverstein RM, Inscoe MN, editors. Behavior-Modifying Chemicals for Insect Management. New York, NY: Marcel Dekker, Inc.; 1990. pp. 437-471. DOI: 10.33338/ef.83484

[67] Hussain A, Hladun S, Vincent M, Wist TJ, Hillier NK, Mori BA. Development of a pheromone monitoring system for the goosefoot groudling moth, *Scrobipalpa atriplicella* (von Röslerstamm) in quinoa, *Chenopodium quinoa* (Wildenow). Crop Protection. 2022;**165**(8):106166. DOI: 10.1016/j.cropro.2022.106166

[68] Boddum T, Skals N, Wiren M, Baur R, Rauscher S, Hillbur Y. Optimisation of the pheromone blend of the swede midge, *Contarinia nasturtii*, for monitoring. Pest Management Science. 2009;**65**(8):851-856. DOI: 10.1002/ps.1762

[69] Narava R, Kumar DVSR, Jaba J, Kumar PA, Rao GVR, Rao VS, et al. Development of temporal model for forecasting of *Helicoverpa armigera* (Noctuidae: Lepidopetra) using arima and artificial neural networks. Journal of Insect Science. 2022;**22**(3):1-9. DOI: https://doi.org/10.1093/jisesa/ieac019

[70] Hartstack AW, Witz JA. Estimating field populations of tobacco budworm moths (Lepidoptera: Noctuidae) from pheromone trap catches. Environmetal Entomology. 1981;**10**:908-914. DOI: 10.1093/ee/10.6.908

[71] Hartstack AW, Witz JA, Buck DR. Moth traps for the tobacco budworm. Journal of Economic Entomology. 1979;**72**:519-522. DOI: 10.1093/jee/72.4.519

[72] Latheef MA, Witz JA, Lopez JD. Relationships among pheromone trap catches of male corn earworm moths (Lepidoptera: Noctuidae), egg numbers, and phenology in corn. The Canadian Entomologist. 1991;**123**:271-281. DOI: 10.4039/Ent123271-2

[73] El-Sayed AM, Suckling DM, Wearing CH, Byers JA. Potential of mass trapping for long-term pest management and eradication of invasive species. Journal of Economic Entomology. 2006;**99**:1550-1564. DOI: 10.1603/0022-0493-99.5.1550

[74] Alpizar D, Fallas M, Oehlschlager AC, Gonzalez LM. Management of *Cosmopolites sordidus* and *Metamasius hemipterus* in banana by pheromone-based mass trapping. Journal of Chemical Ecology. 2012;**38**:245-252. DOI: 10.1007/s10886-012-0091-0

[75] Leskey TC, Bergh JC, Walgenbach JF, Zhang A. Evaluation of pheromone-based management strategies for dogwood borer (Lepidoptera: Sesiidae) in commercial apple orchards. Journal of Economic Entomology. 2009;**102**:1085-1093. DOI: 10.1603/029.102.0329

[76] Yamanaka T. Mating disruption or mass trapping? Numerical simulation analysis of a control strategy for lepidopteran pests. Population Ecology. 2007;**49**:75-86. DOI: 10.1007/s10144-006-0018-0

[77] Allison JD, Johnson CW, Meeker JR, Strom BL, Butler SM. Effect of aerosol surface lubricants on the abundance and richness of selected forest insects captured in multiple-funnel and panel traps. Journal of Economic Entomology. 2011;**104**:1258-1264. DOI: 10.1603/ec11044

[78] Graham EE, Poland TM. Efficacy of fluon conditioning for capturing cerambycid beetles in different trap designs and persistence on panel

traps over time. Journal of Economic Entomology. 2012;**105**:395-401. DOI: 10.1603/EC11432

[79] Dai J, Deng J, Du J. Development of bisexual attractants for diamondback moth, *Plutella xylostella* (Lepidoptera: Plutellidae) based on sex pheromone and host volatiles. Applied Entomology and Zoology. 2008;**43**:631-638. DOI: 10.1303/aez.2008.631

[80] Solomon JD. Trapping and biology of Podesesia and Paranthrene. In Symposium. Pheromones of the Sesiidae, USDA, SEA, Agric Res Rep NE (Washington). 1979; 6:47-54.

[81] Su JW, Zhang GF, Fan WM, Xuan WJ, Sheng CF. The sex pheromone of the rice stem borer, *Chilo suppressalis* in paddy fields: Sticky trap and lure storage time and lure dosage. Chinese Journal of Rice Science. 2001;**15**:197-200 (in Chinese)

[82] Noeth KP, Verleur PM, Bouwer MC, Crous JW, Roux J, Hurley BP, et al. Mass trapping of *Coryphodema tristis* (Lepidoptera: Cossidae) using a sex pheromone in *Eucallytus nitens* compartments in Mpumalanga, South Africa. Journal of Forest Science. 2020;**82**(3):271-279. DOI: 10520/ejc-soufor-v82-n3-a9

[83] Luo Z, Magsi FH, Li Z, Cai X, Bian L, Liu Y, et al. Development and evaluation of sex pheromone mass trapping technology for *Ectropis grisescens*: A potential integrated pest management strategy. Insects. 2020;**11**(1):15. DOI: 10.3390/insects11010015

[84] Al Abassi S, Birkett MA, Pettersson J, Pickett JA, Woodcock CM. Ladybird beetle odour identified and found to be responsible for attraction between adults. Cellular and Molecular Life Sciences: CMLS. 1998;**54**:876-879. DOI: 10.1007/s000180050215

[85] Pinero JC, Dudenhoeffer AP. Mass trapping designs for organic control of the Japanese beetle, *Popillia japonica* (Coleoptera: Scarabaeidae). Pest Management Science. 2018;**74**(7):1687-1693. DOI: 10.1002/ps.4862

[86] Trematerra P. Pheromone use in integrated pest management of stored products. In: Pimentel D, editor. Encyclopedia of Pest Management. Vol. II. Cornell University Ithaca, NewYork, USA: Taylor and Francis Group; 2002. pp. 507-510. DOI: 10.1081/E-EPM-120003841

[87] El-Sayed M, Suckling DM, Byers JA, Jang EB, Wearing CH. Potential of "lure and kill" in long-term pest management and eradication of invasive species. Journal of Economic Entomology 2009;**102**(3):815-835. DOI: https://doi.org/10.1603/029.102.0301

[88] Charmillot PJ, Hofer D, Pasquier D. Attract and kill: A new method for control of the codling moth *Cydia pomonella*. Entomologia Experimentalis et Applicata. 2000;**94**:211-216. DOI: 10.1046/j.1570-7458.2000.00621.x

[89] Suckling DM. Issues affecting the use of pheromones and other semiochemicals in orchards. Crop Protection. 2000;**19**:677-683. DOI: 10.1016/S0261-2194(00)00090-9

[90] Hossain MS, Williams DG, Mansfield C, Bartelt RJ, Callinan L, Il'ichev AL. An attract and kill system to control Carpophilus spp. in Australian stone fruit orchards. Entomologia Experimentalis et Applicata. 2006;**118**:11-19

[91] Tumlinson JH, Guelder RC, Hardee DD, Thompson AC, Hedin PA,

Mınyard JP. Sex pheromones produced by male boll weevil: Isolation, identification and synthesis. Science. 1969;**166**:1010-1012. DOI: 10.1126/science.166.3908.1010

[92] Smith JW. Boll weevil eradication: Area-wide pest management. Annals of the Entomological Society of America. 1998;**91**:239-247. DOI: 10.1093/aesa/91.3.239

[93] Butler SM, Gerry AC, Mullens BA. House fly (Diptera: Muscidae) activity near baits containing (Z)-9-tricosene and efficacy of commercial toxic fly baits on a Southern California dairy. Journal of Economic Entomology. 2007;**100**:1489-1495. DOI: 10.1093/jee/100.4.1489

[94] Geden CJ, Szumlas D, Walker TW. Evaluation of commercial and field-expedient baited traps for house flies, *Musca domestica* L. (Diptera: Muscidae). Journal of Vector Ecology. 2009;**34**:99-103. DOI: 10.1111/j.1948-7134.2009.00012.x

[95] Witzgall P, Kirsch P, Cork A. Sex pheromones and their impact on pest management. Journal of Chemical Ecology. 2010;**36**:80-100. DOI: 10.1007/s10886-009-9737-y

[96] Butler GD Jr, Las AS. Predaceous insects: Effect of adding permethrin to the sticker used in gossyplure applications. Journal of Economic Entomology. 1983;**76**:1448-1451

[97] Beasley CA, Henneberry TJ. Combining gossyplure and insecticides in pink bollworm control. California Agriculture. 1984;**38**:22-24

[98] Downham MCA, McVeigh LJ, Moawad GM. Field investigation of an attracticide control technique using the sex pheromone of the Egyptian cotton leafworm, *Spodoptera littoralis* (Lepidoptera: Noctuidae). Bulletin of Entomological Research. 1995;**85**:463-472

[99] Trematerra P. The use of sex pheromones to control *Ephestia kuehniella* Zeller (Mediterranean flour moth) in flour mills by mass trapping and attracticide (lure and kill) methods. In Highley E, Wright EJ, Banks HJ and Champ BR, editors, Proceedings of the 6th Working Conference on Stored product Protection Canberra, Australia, 375-382.

[100] Hossain MS, Williams DG, Mansfield C, Bartelt RJ, Callinon L, Il'ichev AL. An attact and kill system to control Carpophilus spp. in Australian stone fruit orchards. Entomologie Experimentalis et Application. 2006;**118**(1):11-19. DOI: 10.1111/j.1570-7458.2006.00354.x

[101] Carde RT. Principles of Mating Disruption. New York, NY, USA: Marcel Dekker; 1990

[102] Carde RT, Minks AK. Control of moth pests by mating disruption: Successes and constraints. Annual Review of Entomology. 1995;**40**:559-585. DOI: 10.1146/annurev.en.40.010195.003015

[103] Millar JG. Insect pheromones for integrated pest management: Promise versus reality. Redia. 2007;**90**:51-55

[104] Benelli G, Lucchi A, Thomson D, Ioriatti C. Sex pheromone aerosol devices for mating disruption: Challenges for a brighter future. Insects. 2019;**10**:308. DOI: 10.3390/insects10100308

[105] Barclay HJ, Judd GJR. Models for mating disruption by means of pheromone for insect pest control. Researches on Population Ecology.

1995;**37**(2):239-247. DOI: 10.1007/BF02515826

[106] Mafra-Neto A, Fettig CJ, Unson AS, Rodriguez-Saona C, Holdcraft R, Faleiro JR, et al. Chapter 15: Development of specialized pheromone and lure application technologies (SPLAT®) for management of Coleopteran pests in agricultural and forest systems. In: Gross AD, Coats JR, Duke SO, Seiber JN, editors. Biopesticides: State of the Art and Future Opportunities, ACS Symposium Series. Vol. Vol. 1172. Washington, DC: American Chemical Society; 2014. pp. 214-242. DOI: 10.1021/bk–2014-1172.ch015

[107] Zhang XZ. Discovery of *Cydia pomonella* in China. Acta Entomologica Sinica. 1957;7:467-472

[108] Cui G, Zhu JJ. Pheromone–based pest management in China: Past, present and future prospects. Journal of Chemical Ecology. 2016;**42**:557-570. DOI: 10.1007/s10886-016-0731-x

[109] Lance DR, Leonard DS, Mastro VC, Walters ML. Mating disruption as a suppression tactic in programs targeting regulated lepidopteran pests in US. Journal of Chemical Ecology. 2016;**42**:590-605. DOI: 10.1007/s10886-016-0732-9

[110] Witzgall P, Stelinski L, Gut L, Thomson D. Codling moth management and chemical ecology. Annual Review of Entomology. 2008;**53**:503-522. DOI: 10.1146/annurev.ento.53.103106.093323

[111] Gordon D, Zahavi T, Anshelevich L, Harel M, Ovadia S, Dunkelblum E, et al. Mating disruption of *Lobesia botrana* (Lepidoptera: Tortricidae): Effect of pheromone formulations and concentrations. Journal of Economic Entomology. 2005;**98**:135-142. DOI: 10.1093/jee/98.1.135

[112] Stelinski LL, Miller JR, Ledebuhr R, Siegert P, Gut LJ. Season-long mating disruption of *Grapholita molesta* (Lepidoptera: Tortricidae) by one machine application of pheromone in wax drops (SPLAT-OFM). Journal of Pest Science. 2007;**80**:109-117. DOI: 10.1007/s10340-007-0162-0

[113] Trematerra P, Athanassiou C, Stejskal V, Sciarretta A, Kavallieratos N, Palyvos N. Large-scale mating disruption of Ephestia spp. and *Plodia interpunctella* in Czech Republic, Greece and Italy. Journal of Applied Entomology. 2011;**135**:749-762. DOI: 10.1111/j.1439-0418.2011.01632.x

[114] Hoshi H, Takabe M, Nakamuta K. Mating disruption of a carpenter moth, *Cossus insularis* (Lepidoptera: Cossidae) in apple orchards with synthetic sex pheromone, and registration of the pheromone as an agrochemical. Journal of Chemical Ecology. 2016;**42**:606-611. DOI: 10.1007/s10886-016-0723-x

[115] Il'ichev AL. Area-wide application of pheromone mediated mating disruption in sustainable IPM. IOBC WPRS Bulletin. 2006;**29**:95-104. DOI: 10.13140/2.1.1339.0722

[116] Ünlü L, Özgür E, Uulu TE. Control of *Tuta absoluta* (Meyrick) (Lepidoptera: Gelechiidae) in open fşeld tomato crops using the mating disruption technique. Phytooparasitica. 2021;**49**:385-396. DOI: 10.1007/s12600-021-00881-4

[117] Liang Y, Luo M, Fu X, Zheng L, Wei H. Mating disruption of *Chilo suppressalis* from sex pheromone of another pyralid rice pest *Cnaphalocrocis medinalis* (Lepidoptera: Pyralidae). Journal of Insect Science. 2020;**20**(3):19. DOI: 10.1093/jisesa/ieaa050

[118] Ferracini C, Pogolotti C, Rama F, Lentini G, Saitta V, Mereghetti P, et al.

Pheromone-mediated mating disruption as management option for *Cydia* spp. in Chestnut Orchard. Insects. 2021;**12**(10):905. DOI: 10.3390/insects12100905

[119] Lucchi A, Suma P, Ladurner E, Iodice A, Savino F, Ricciardi R, et al. Managing the vine mealybug, *Planococcus ficus*, through pheromone-mediated mating disruption. Environmental Science and Pollution Research. 2019;**26**:10708-10718. DOI: 10.1007/s11356-019-04530-6

[120] Maki EC, Millar JG, Rodstein J, Hanks LM, Barbour JD. Evaluation of mass trapping and mating disruption for managing *Prionus californicus* (Coleoptera: Cerambycidae) in hop production yards. Journal of Economic Entomology. 2011;**104**:933-938. DOI: 10.1603/EC10454

[121] Wenninger EJ, Averill AL. Mating disruption of oriental beetle (Coleoptera: Scarabaeidae) in cranberry using retrievable, point-source dispensers of sex pheromone. Environmental Entomology. 2006;**35**(2):458-464. DOI: 10.1603/0046-225X-35.2.458

[122] Wijayaratne LKW, Burks CS. Persistence of mating suppression of the Indian meal moth *Plodia interpunctella* in the presence and absence of commercial mating disruption dispensers. Insects 2020; 11(10):701. https://doi.org/10.3390/insects11100701

[123] Gavara A, Navarro-Llopis V, Primo J, Vacas S. Influence of weather conditions on *Lobesia botrana* (Lepidoptera: Tortricidae) mating disruption dispensers' emission rates and efficacy. Crop Protection. 2022;**155**:105926. DOI: 10.1016/j.cropro.2022.105926

[124] Gut LJ, Brunner JF. Pheromone-based management of codling moth (Lepidoptera: Tortricidae) in Washington apple orchards. Journal of Agricultural Entomology. 1998; 15:387-406. ISSN: 0735-939X.

[125] Trimble RM. Mating disruption for controlling the codling moth, *Cydia pomonella* (L) (Lepidoptera: Tortricidae), in organic apple production in southwestern Ontario. The Canadian Entomologist. 1995;**127**(4):493-505. DOI: 10.4039/Ent127493-4

[126] Vickers RA, Thwaite WG, Williams DG, Nicholas AH. Control of codling moth in small plots by mating disruption: Alone and with limited insecticide. Entomologia Experimentalis et Applicata. 1998;**86**:229-239. DOI: 10.1046/j.1570-7458.1998.00285.x

[127] Webb RE, Leonhardt BA, Plimmer JR, Tatman KM, Boyd VK, Coehn DL, et al. Effect of racemic disparlure released from grids of plastic ropes on mating success of gypsy moth (Lepidoptera: Lymantriidae) as influenced by dose and by population-density. Journal of Economic Entomology. 1990;**83**(3):910-916. DOI: 10.1093/jee/83.3.910

[128] Borchert DM, Walgenbach JF. Comparison of pheromone-mediated mating disruption and conventional insecticides for management of tufted apple bud moth (Lepidoptera: Tortricidae). Journal of Economic Entomology. 2000;**93**(3):769-776. DOI: 10.1603/0022-0493-93.3.769 PMID: 10902329

[129] Ansebo L, Ignell R, Lofqvist J, Hansson BS. Responses to sex pheromone and plant odours by oldfactory seceptor reurons housed in sensilla auricillica of the codling moth, *Cydia pomonella* (Lepidoptera: Tortricidae). Journal of Insect Physiology. 2005;**51**:1066-1074. DOI: 10.1016/j.jinsphys.2005.05.003

[130] Knight AL, Light DM. Monitoring codling moth (Lepidoptera: Tortricidae)in sex-pheromone treated orchards with (E)-4,8-dimethyl-1,3,7-nonatriene or pear ester in combination with codlemone and acetic acid. Environmental Entomology. 2012;**41**:407-414. DOI: 10.1603/EN11310

[131] Knight AL, Stelinski LL, Hebert V, Gut L, Light D, Brunner J. Evaluation of novel semiochemicals dispensers simultaneously releasing pear ester and sex pheromone for mating disruption of codling moth (Lepidoptera: Tortricidae). Journal of Applied Entomology. 2012;**136**:79-86. DOI: 10.1111/j.1439-0418.2011.01633.x

[132] Stelinski LL, Gut LJ, Miller JR. An attempt to increase efficacy of moth mating disruption by co-releasing pheromone with kairomones and to understand possible underlying mechanisms of this technique. Environmental Entomology. 2013;**42**:158-166. DOI: 10.1603/EN12257

[133] Onufrieva KS, Thorpe KW, Hickman AD, Leonard DS, Mastro VC, Roberts EA. Gypsy moth mating disruption in open landscapes. Journal of Agricultural Forest Entomology. 2008;**10**:175-179. DOI: 10.1111/j.1461-9563.2008.00375.x

[134] Sharov AA, Leonard D, Liebhold AM, Clemens NS. Evaluation of preventive treatments in low-density gypsy moth populations using pheromone traps. Journal of Economic Entomology. 2002;**95**:1205-1215. DOI: 10.1603/0022-0493-95.6.1205

[135] Antonaroli R. Control of two tortricid chestnut pests by mating disruption. Informatore Agrario. 2000;**56**:89-91

[136] Hummel HE, Lagnner SS, Eisinger MT. Pheromone dispensers, including organic polimer fibers, described in the crop protection literature: Comparasion of their innovation potential. Communications in Agricultural and Applied Biological Sciences. 2013;**78**:233-252 PMID: 25145244

[137] Hummel HE, Langner SS, Breuer M. Electrospun mesofibers, a novel biodegredable pheromone dispenser technology, are combined with mechanical deployment for efficient IPM of *Lobesia botrana* in vineyards. Communications in Agricultural and Applied Biological Sciences. 2015;**80**:331-341 PMID: 27141730

[138] Lucchi A, Ladurner E, Iodice A, Savino F, Ricciardi R, Cosci F, et al. Eco-friendly pheromone dispensers-a green route to manage the European grapevine moth? Environmental Science and Pollution Research. 2018;**25**:9426-9442. DOI: 10.1007/s11356-018-1248-3

[139] Pyke B, Rice M, Sabine B, Zalucki MP. The push-pull strategy behavioral control of Heliothis. Australian Cotton Growers. 1987;**9**:7-9

[140] Miller JR, Cowles RS. Stimulo-deterrent diversion: A concept and its possible application to onion maggot control. Journal of Chemical Ecology. 1990;**16**(11):3197-3212. DOI: 10.1007/BF00979619

[141] Khan ZR, Pickett JA. The 'Push–Pull' strategy for stem borer management: A case study in exploiting biodiversity and chemical ecology. In: Gurr G, Waratten SD, Altieri MA, editors. Ecological Engineering for Pest Management: Advances in Habitat Manipulations for Arthropods. Oxfordshire, UK: CABI Publishing; 2004. pp. 155-164

[142] Cook SM, Khan ZR, Pickett JA. The use of push-pull strategies in integrated

pest management. Annual Review of Entomology. 2007;**52**:375-400. DOI: https://doi.org/10.1146/annurev.ento.52.110405.091407

[143] Khan ZR, Midega CAO, Bruce TJA, Hooper AM, Pickett JA. Exploiting phytochemicals for developing a 'push-pull' crop protection strategy for cereal farmers in Africa. Journal of Experimental Botany. 2010;**61**(15):4185-4196. DOI: 10.1093/jxb/erq229

[144] França de SM, Mariana OP, Cesar AB, José VO. The use of behavioral manipulation techniques on synthetic insecticides optimization. In: Turdan S editors. Insecticides – Development of Safer and More Effective Technologies. InTech Design Team, Croatia, 2013; p. 177-196.DOI: https://doi.org/10.5772/53354

[145] Abd El-Ghany NM, Abdel-Razek AS, Djelouah K, Moussa A. Efficacy of some eco-friendly biopesticides against *Tuta absoluta* (Meyrick). Bioscience Research. 2018;**15**(1):28-40

[146] Smart LE, Blight MM, Pickett JA, Pye BJ. Development of field strategies incorporating semiochemicals for the control of the pea and bean weevil, *Sitona lineatus* L. Crop Protection. 1994;**13**:127-135. DOI: 10.1016/0261-2194(94)90163-5

[147] Martel JW, Alford AR, Dickens JC. Synthetic host volatiles increase efficacy of trap cropping for management of Colorado potato beetle, *Leptinotarsa decemlineata* (Say). Agriculture for Entomology. 2005;7:79-86. DOI: 10.1111/j.1461-9555.2005.00248.x

[148] Mauchline AL, Osborne JL, Martin AP, Poppy GM, Powel W. The effects of non-host plant essential oil volatiles on the behavior of the pollen beetle *Meligethes aeneus*. Entomologia Experimentalis et Applicata. 2005;**114**:181-188. DOI: 10.1111/j.1570-7458.2005.00237

[149] Wallinford AK, Cha DH, Loeb GM. Evaluating a push-pull strategy for management of *Drosophila suzukii* Matsumura in red raspberry. Pest Management Science. 2018;**74**:120-125. DOI: 10.1002/ps.4666

[150] Midega CAO, Pittchar JO, Pickett JA, Hailu GW, Khan ZR. A climate-adapted push-pull system effectively controls fall armyworm, *Spodoptera frugiperda* (J E Smith), in maize in East Africa. Crop Protection. 2018;**105**:10-15. DOI: 10.1016/j.cropro.2017.11.003

[151] Balaso GM, Nalini C, Yankit P, Thakur P. Push-pull strategy: Novel approach of pest management. Journal of Entomology and Zoology Studies. 2019;7(5):220-223

[152] Chatterjee D, Kundu A. Push pull strategy of integrated pest management. Just Agriculture. 2022; 2(9):1-6. ISSN: 2582-8223

[153] Powell W, Pickett JA. Manipulation of parasitoids for aphid pest management: Progress and prospects. Pest Management Science. 2003;**59**(2):149-155. DOI: 10.1002/ps.550

[154] Ross DW, Daterman GE. Reduction of Douglas-fir beetle infestation of high-risk stands by antiaggregation and aggregation pheromones. Canadian Journal of Forest Research. 1994;**24**:2184-2190. DOI: 10.1139/x94-281

[155] Bentz BJ, Kegley S, Gibson K, Thier R. A test of high-dose verbenone for stand-level protection of lodgepole and whitebark pine from mountain pine beetle (Coleoptera: Curculionidae: Scolytinae) attacks.

Journal of Aconomic Entomology. 2005;**98**:1614-1621

[156] Borden JH, Birmingham AL, Burleigh JS. Evaluation of the push pull tactic against the mountain pine beetle using verbenone and non-host volatiles in combination with pheromone-baited trees. The Forestry Chronicle. 2006;**82**:579-590. DOI: 10.5558/tfc82579-4

[157] Gillette NE, Hansen EM, Mehmel CJ, Mori SR, Webster JN, Erbilgin N, et al. Area-wide application of verbenone-releasing flakes reduces mortality of whitebark pine *Pinus albicaulis* caused by the mountain pine beetle *Dendroctonus ponderosae*. Agricultural and Forest Entomology. 2012;**14**:367-375. DOI: 10.1111/j.1461-9563.2012.00577.x

[158] Gilette NE, Mehmet CJ, Mori SR, Webster JN, Wood DL, Erbilgin N, et al. The push-pull tactic for mitigation of mountain pine beetle (Coleoptera: Curculionidae) damage in lodgepole and whitebark pines. Environmental Entomology. 2012;**41**:1575-1586. DOI: 10.1603/EN11315

[159] Progar RA. Five year operational trial of verbenone to deter mountain pine beetle (*Dendroctonus ponderosae*.i Coleoptera:Scolytidae) attack of lodgepole pine (*Pinus contorta*). Environmental Entomology. 2005;**34**:1402-1407. DOI: 10.1603/0046-225X-34.6.1402

[160] Giblin-Davis RM, Gries R, Crespi B, Robertson LN, Hara AH, Gries G, et al. Aggregation pheromones of two geographical isolates of the New Guinea sugarcane weevil, *Rhabdoscelus obscurus*. Journal of Chemical Ecology. 2000;**26**:2763-2780. DOI: 10.1023/A:1026

Chapter 4

Perspective Chapter: Tapping Significance of Microbial Surfactants as a Biopesticide and Synthetic Pesticide Remediator – An Ecofriendly Approach for Maintaining the Environmental Sustainability

Shikha Gaikwad

Abstract

Pests are the major concern for plant infections that affect the agriculture production drastically and result in the loss of economy. Regular use of synthetic chemicals develops resistance in pests and affects soil, plant, and human health. The development and promotion of green technology facilitated by microbiota helps in maintaining a healthy environment. Molecules of microbial origin are well-known elicitors for stimulating and sustaining the plant immune system and fertility level of the soil. They compete with the pathogens for resources like food and space, activate the inherent defenses of plants, synthesize antimicrobial chemicals, or other metabolites that degrade and remediate synthetic chemicals. Biosurfactant (BS) is an important amphiphilic molecule with polar and non-polar. Its structure contributes to its high biodegradability, low toxicity, and stability in harsh environments. In the chapter, the multifunctional properties of biosurfactants, methods used for evaluating their biosurfactant producing abilities, methods used for identification, and characterization of the chemical structure of biosurfactants, along with the significance of metagenomics documented. The mechanisms of biosurfactants in controlling the growth of pests and their importance as pesticide remediations are also discussed.

Keywords: biosurfactants, biocontrol agent, biopesticide, pesticide remediator, metagenomics techniques

1. Introduction

All countries have concern about meeting the expanding food needs, which places an additional burden on the agricultural and food business. Controlling the spread of phytopathogens and raising the soil's fertility level have been top priorities in efforts to increase crop output. To cater to ever-increasing demand, chemical fertilizers, insecticides, and fungicides are continuously used for decades. Pesticides at the rate of 2 million tons are used annually, with 2% being rodenticides and nematicides and 50% being herbicides, insecticides, and fungicides [1]. These pesticide residues endure in the water and soil for years and develop pesticide resistance among deleterious microbes [2–4]. It has been reported that milk, meat, and other food products contain considerable quantities of pesticide residues [5]. Boedeker with his coworkers highlighted, harmful effects of synthetic chemicals on almost 44% of the farming population every year [6]. The most prevalent morbidities that are connected to these practices include immune system deficits, pulmonary dysfunction, and malignancy [7, 8]. The primary cause of these consequences may be the overuse of agrochemicals, which has been a considerable issue [9]. To enhance soil fertility and prevent insect infestation, it is necessary to consider the revitalization of native soil systems that will resist the use of these synthetic chemical amendments. Finding solutions that hold sustainability in the environment is required urgently. Numerous physical, chemical, and biological strategies are frequently used to solubilize and/or degrade hazardous substances. The use of membrane filtration, adsorption, soil washing, granular activated carbon, and photocatalytic remediation are some of the common examples of physical approaches. Ion-exchange, precipitation, coagulation, floatation, and flocculation procedures are examples of chemical-based methods. These physicochemical techniques that are in use at frequent intervals are relatively futile and unsustainable. Whereas combinations of physicochemical techniques occasionally prosper [10]. It is reported that synthetic surfactants are combined in many pesticide formulations as adjuvants [11]. Synthetic surfactants are expensive, caustic, and impervious to degradation. Hence a green approach that conforms to the strict guidelines of green chemistry and green technology, which are in high demand in the modern period, is essential. It will support to maintain and sustain the desired healthy flora and fauna, which can deliberately keep the ecosystem healthy.

2. Overview of biosurfactants

Surface-active substances known as biosurfactants (BS) are synthesized by some of the prokaryotic and eukaryotic microorganisms in their stationary stage of development. According to a recent survey by Global Market Insight, between 2020 and 2026 year, the BS market is anticipated to expand at a rate of more than 5.5% Compound Annual Growth Rate. In 2019, the market for BS was worth more than USD 1.5 billion [12]. Microorganisms generate BS either extracellularly or as an ingredient of the cell membrane [13, 14]. They are prevalent because of their wide range of benefits against synthetic surfactants (**Figure 1**).

BS nature is highly dependent on the microbial origin and on the available nutrients. In accordance with their chemical complexity or molecular weight, they are categorized as high and low molecular weight BS. Their molecular weights range from 50 to 1500 kDa LWM-BS (Low Molecular Weight Biosurfactants) decrease the surface and interfacial tensions at the air and water boundaries. In disparity, HMW-BS (High Molecular Weight Biosurfactants) also described as "bioemulsans," are efficient at stabilizing the oil in water emulsions (**Table 1**).

Based on the chemical structure, it is an amphiphilic molecule, with a polar head and nonpolar tail. The hydrophilic moiety is made up of intrinsically simple ester, phosphate, hydroxyl, or carboxyl groups, as well as carbohydrates like

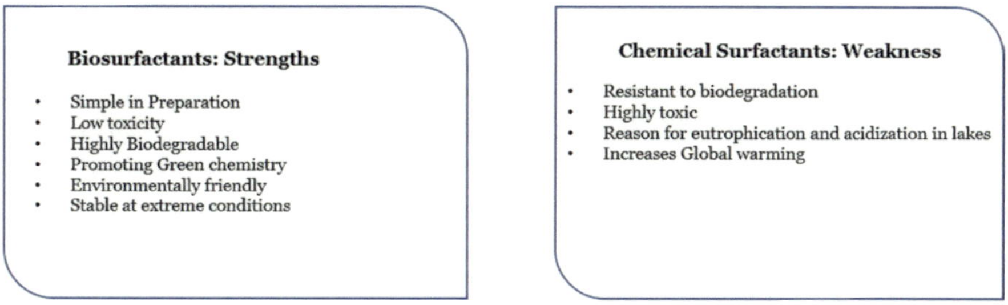

Figure 1.
Biosurfactants strengths and weakness of chemical surfactants.

Biosurfactants	Bioemulsifiers
Low molecular weight-compounds	High molecular weight- compounds
Reduces surface tension and emulsification	The emulsion formed are not stable
Used for stabilization of emulsion	Used for only emulsification
For example-lipopeptides, glycolipids (Rhamnolipids, Sophorolipids, Trehalose lipids, Phospholipids, Corinomiocolic acid, fatty acids), Lipopeptides (Surfactin, Wincnsin, Gramicidin, Substilsin, Peptide lipid, Lichenysin)	For example-polysaccharides, (Emulsan, Biodispersion, Mannan-lipid protein, Carbohydrate lipid-, protein), Particulate (Vesicles)

Table 1.
Differentiating points of biosurfactants and bioemulsifiers.

monosaccharides, oligosaccharides, and polysaccharides, as well as proteins, amino acids, and peptides. Unsaturated or saturated fatty acids, hydroxyl fatty acids, or fatty alcohols make up the hydrophobic portion (**Figure 2**).

Based on its structural characteristics, BS has several advantageous properties which enable them to be employed as an adjuvant or essential component in a wide range of formulations, as a potential bioremediator for synthetic chemicals, and as an emulsifier in an eclectic assortment of industrial applications, such as those involving food and beverage, petroleum, cosmetics, organic chemicals, and

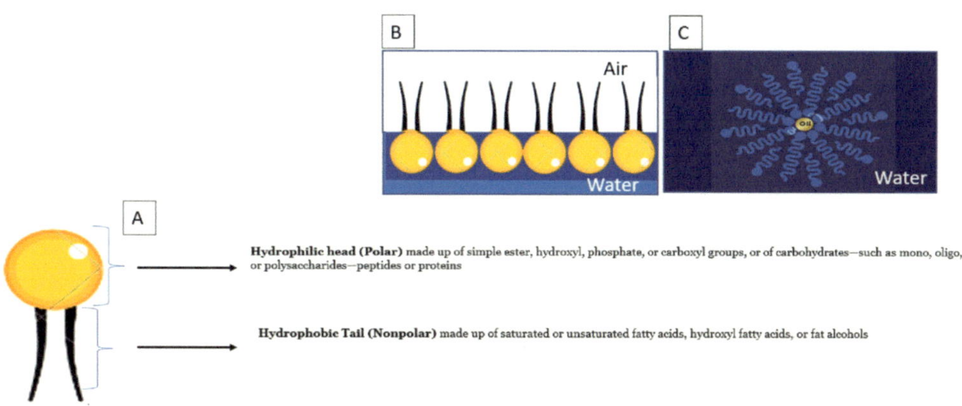

Figure 2.
(A) An amphipathic structure of a biosurfactant, (B) surfactant monomer at the surface of water and (C) micelle formation above CMC (critical micelle concentration).

pharmaceuticals [15]. BS occupies a unique place in the agricultural sector due to its capacity of emulsification, dispersion, solubilization, foaming, and wetting agents, which accelerate hydrophobic molecules' solvation in aqueous media for the formation of emulsion [16–18]. They have proved themselves to be a top contender for leading the way in the field of agricultural and environmental science [19–28].

3. Multifunctional properties of biosurfactants

BSs' amphipathic structure is giving them a special place to showcase their multifunctional properties [16]. A few important multidimensional properties are mentioned in **Figure 3** and detailed description is listed below:

3.1 Surface and interfacial activity

Surface tension is a phenomenon that happens when a liquid surface and another phase come together (it can be a liquid as well). The molecules at a liquid's surface are drawn toward the liquid's center by the contracting force of surface tension. The surface and interfacial tensions of certain fluids can be reduced by BS at extremely low concentrations due to lower Critical Micelle Concentrations (CMC).

Figure 3.
Functional properties of biosurfactants.

3.2 Solubilization

In the aqueous phase, biosurfactant molecules spontaneously group together and form supramolecular nano sized aggregates with a hydrophobic core and a hydrophilic surface exposed to water. It is formed when BS concentration surpasses a threshold value which is well known as critical micelle concentration—CMC. These small sized aggregations enhance the bioavailability of water-insoluble substances in aqueous liquids or chemical agents by allowing the transportation and confinement of insoluble molecules.

3.3 Wetting

It is the capability of a liquid to provide a link with another surface and to spread out evenly and easily over the top of the surface of another solid or fluid. When a liquid with a high wetting capacity spread across a surface, it creates a thin, continuous film. Biosurfactants are effective wetting agents because they lower liquid surface tension by reducing attractive forces, which enhances affinity toward different surfaces. Instead of connecting them to the surface tension, it penetrates the pores.

3.4 Foaming

Small levels of BS can decrease a liquid's surface tension (lessen the effort required to produce foam) or boost its colloidal stability by preventing bubble coalescence. Biosurfactants are intense at the gas-liquid interface, where they form fizzes that move through the liquid and produce foam. BS is a substance that encourages the production of foam.

3.5 Dispersion

Dispersion results from the decrease in the cohesive attraction between similar particles. A few BS are applied as a dispersion agent to reduce the aggregation of insoluble particles amid each other in the suspension. For example, BS removes hydrophobic molecules from the surfaces of rocks to improve their mobility and recovery in oil extraction steps. Dispersion also plays an additional role to reduce or completely eradicate the biofilm formation of undesirable microbes.

3.6 Flocculation

The process of flocculation starts when colloidal particles, either natively or because of the addition of a clarifying agent, migrate from a condition of suspension to sediment in the form of floc or flake. Emulsions are restored to their original state by these flocs, which can be distorted by mechanical force and aren't permanent.

3.7 Emulsification

It is the process of combining liquids that ordinarily do not mix to create an emulsion. BS has emulsifying and demulsifying properties. The two most prevalent emulsions are water-in-oil (w-0) and oil-in-water (o-w). They are typically unstable in two-phase solutions. Biosurfactants signifies the solubilization of large particles with micellar structures by assisting the dispersion of one liquid into another and making it easier for two immiscible liquids to be mixed.

3.8 De-emulsification

In this process, the stable interface between the internal and bulk levels gets disturbed, due to which the emulsions get split. BS makes the process of de-emulsification easier.

3.9 Biodegradability

BS is a metabolic byproduct of microorganisms and is easily broken down in nature without generating any deleterious byproducts.

3.10 Biocompatibility, low toxicity, and digestibility

They are not harmful or toxic to living tissue. When they are interacting with other organisms, they do not change their bioactivity or mechanism.

3.11 Stability at various extreme conditions

Several BS stay stable across a diversity of adverse environmental situations, such as pH, temperature, and salinity, rendering them potential candidates to be deployed in a comprehensive array of industries.

4. Methods to detect biosurfactants producing ability of a microbial strain

To choose biosurfactant synthesizers from a set of microorganisms, various screening techniques are advocated concurrently (**Figure 4**). A single evaluation system is never recommended [29], various techniques that can be used are:

4.1 Surface tension and interfacial measurement

The aspects of surfaces and interfaces, such as the surface excess concentration, adsorption kinetics, surface pressure, and CMC can be recognized with the aid of tensiometers. These instruments differ in terms of the underlying physical concepts, the mechanical layout, the type of measurements they can perform—static or dynamic—and whether they can detect surface tension (ST), interfacial tension (IM), or both. It is possible to follow the adsorption kinetics by making dynamic measurements on surfaces or interfaces that are not in equilibrium. The most common techniques that are used to evaluate ST, IM or both are Du Noüy ring, Wilhelmy plate, and pendant drop. The strength required to separate a wire ring or loop from a surface or interface is the cornerstone for these strategies. The detachment force and interfacial tension have an inverse relationship. The accuracy, simplicity, and minimal sample volume are needed to make this approach more advantageous. It is the easiest screening technique and is suitable for a preliminary screening.

4.2 Drop collapse method

Polar water molecules are attracted away from the hydrophobic surface in the lack of surfactants, which helps to stabilize the drops. In comparison, when there are surfactants in the liquid, the drops spread or even collapse due to the reduced force

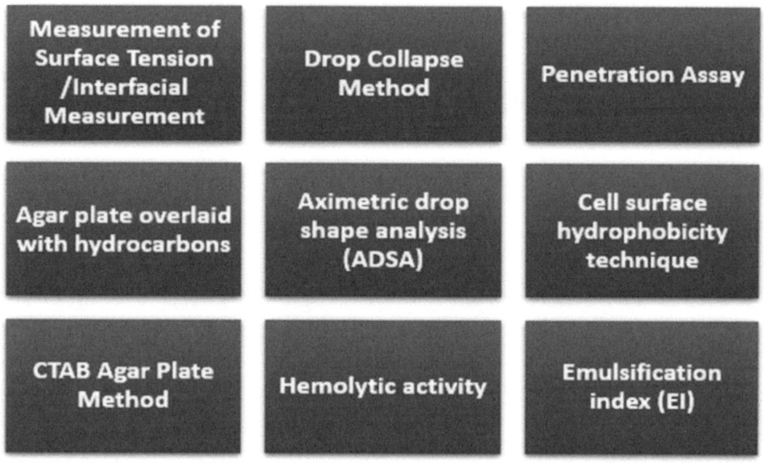

Figure 4.
Methods to detect biosurfactants producing ability of microbes.

or interfacial tension. Drop stability is correlated with surface and interfacial tension and is reliant on surfactant content. It is easy, quick, and simple, and needs a little volume of sample and no specific equipment.

4.3 Penetration assay

This test requires contact between two insoluble phases to function. A staining solution is applied to culture supernatants for easy differentiation. If BS is present, the colored supernatant passes through the oil film barrier and enters the paste. Within 15 min, it transforms the color to murky white. Supernatant devoid of biosurfactants would lose its color when becoming cloudy. It is a quick and effective method to screen through many potential isolates.

4.4 Agar plate overlaid with hydrocarbons

In an oil coated agar plate, an axenic culture is streaked and incubated for 5–7 days at a desired temperature. An emulsified halo around the streaked colonies can be identified as BS producers.

4.5 Aximetric drop shape analysis (ADSA)

The technique includes assessing the drop's shape, which is strongly impacted by the equilibrium between surface tension and outside forces, like gravity. The shape of the drop exemplifies its range of utility and validity.

4.6 Cell surface hydrophobicity technique

It is based on a technique that employs microbes to attach to hydrocarbons. Microorganisms that can actively absorb hydrocarbons typically have highly hydrophobic coverings.

4.7 CTAB (cetyltrimethylammonium bromide) agar plate method

It is used basically to identify their anionic, cationic, nonionic, and neutral nature. The media is supplemented with different dyes such as cetyltrimethylammonium bromide and methylene blue. Presence and absence of a halo zone surrounding the culture reflects its nature and property of the BS.

4.8 Hemolytic activity

In media such as Luria agar (LA) or nutrient agar (NA) fresh blood is mixed. Microbe is streaked and incubated at the desired temperature. After incubation hemolysis zone was observed around the colony which indicates BS synthesizing ability of microbes.

4.9 Emulsification index (EI)

Emulsification activity is calculated by EI. Hydrocarbon is added to the culture broth and mixed. Subsequently allowed to stand overnight. The height of the emulsion formed between aqueous, and hydrocarbon is measured, that defines the stability and strength of a surfactant.

5. Methods to identify and characterize biosurfactants

New approaches are evolving for recognizing and identifying BS because of scientific and technological innovations. The techniques that are used to identify and characterize the BS are as follows (**Figure 5**).

5.1 Thin layer chromatography (TLC)

The most popular, straightforward, and affordable approach for identifying various macromolecules, including carbohydrates, lipids, fats, proteins, amino acids, and peptides. In this methodology, an adsorbent medium such as aluminum oxide (alumina), silica gel, or cellulose is wrapped thinly around a sheet of glass, plastic, or glass. The sample is placed on a plate and given time to go through several mobility phases. The development of the colors after spraying developers helps in recognizing the presence of macromolecules that help in identifying the class and subclass of BS.

5.2 High performance-liquid chromatography (HPLC)

It is the most comprehensive and efficient quantitative tactic. A detector, a mobile phase, and a fixed phase constitute it. The mobile phase moves the sample solution once it is introduced through the injector port. The mobile phase progressively lowers the components of the sample solution over the stationary phase, which is a solid. The migration of the components is regulated by the noncovalent interactions between the compound and the column. This methodology considers several components according to their polarity. The segregated products may then be recognized, and fractions for certain peaks can be obtained to investigate the structure of BS moiety.

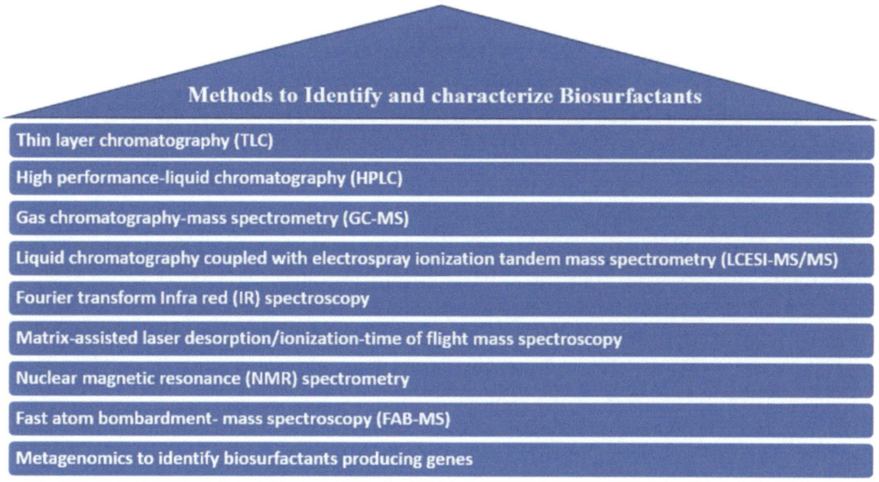

Figure 5.
Methods to identify and characterize biosurfactants.

5.3 Gas chromatography-mass spectrometry (GC-MS)

Mass spectrometry, coupled with either gas or liquid chromatography, is utilized to discover the chemical bonds and structure of BS. In addition, the procedure provides a qualitative and quantitative approach used for detecting the chemical nature of BS. The mass spectrometers are made up of three important components: an ion source, a molecular mass analyzer, and a detector. As a result, volatile samples are directly introduced into the device, whereas non-volatile samples need to be dissolved in volatile solvents. As the sample passes through the electromagnetic field, it immediately becomes ionized. The computer system receives the amplified automated signal and reports it as various chromatogram peaks. It is therefore essential to evaluate the overall quantity and quality of the compounds. The BS compound's hydrophobic component (water-repelling) is typically revealed by GC-MS.

5.4 Liquid chromatography coupled with electro-spray ionization tandem mass spectrometry (LCESI-MS/MS)

The structural makeup of the hydrophilic moiety is revealed by LC-MS (water-loving). Different research centers are currently using these approaches to detect BS biomolecules, which reduces the chance of erroneous characterization. It is defined as a less expensive and less time-consuming procedure and detects BS even at low concentrations.

5.5 Fourier transform infra-red (IR) spectroscopy

It is the most typical type of infrared spectroscopy. According to the underlying hypothesis, some infrared (IR) light is absorbed when it travels through a sample. Molecules with covalent bonds will selectively absorb different wavelengths of light, affecting the bond's vibrational energy. The type of vibration (stretching or bending) depends on the atoms in the linkage. The quantity of radiation that hits the sample is calculated. The spectrum is represented by plotting transmittance and wavenumber (cm^{-1}). It is utilized in the range of around 4000 and 400 cm^{-1}. It detects and discriminates the spectra between molecules. The sample is not destroyed, and it is substantially faster, more sensitive with an inexpensive strategy.

5.6 Matrix-assisted laser desorption/ionization-time of flight mass spectroscopy (MALDI-TOF)

This method is based on a laser energy-absorbing matrix that creates ions from large molecules with little fragmentation. The MALDI approach is a three-step procedure. The sample is first applied to a metal plate after being blended with a suitable matrix material. After that, the sample is exposed to a pulsed laser, which detects the sample and matrix material. The analyte molecules are then accelerated into a certain mass spectrometer that is being utilized to examine them. After detection, molecules are finally ionized by protonation or deprotonation in the hot plume of gases. Although the process is robust, its reliability is offering its special place in the analytical world. It has been used to analyze molecules like DNA, proteins, peptides, carbohydrates, polymers, and dendrimers that have a propensity to be brittle and fragment when ionized by more conventional ionization techniques.

5.7 Nuclear magnetic resonance (NMR)-spectrometry

It is constructed on transformations that emerge in magnetically significant atoms when an outside magnetic field exists. In a process nucleus absorbs radio frequency radiation, due to which the nuclear spin realigns or splits in the higher-energy direction and emits radiation again and returns the molecule again to the lower-energy state. For each nucleus, the magnetogyric ratio serves as a proportionality factor. NMR can be used to identify the functional groups and linkages within lipid and carbohydrate molecules. The specific location of each functional group and facts about the structural isomers can be revealed by a series of NMR spectroscopy.

5.8 Fast atom bombardment-mass spectroscopy (FAB-MS)

It relies on ionization from the liquid phase, with the probable requirement that the sample molecules congregate at the surface of the liquid matrix in the vacuum. It highlights the BS's lipid structure.

5.9 Metagenomics to identify biosurfactants producing genes

"Metagenomics" is a combination of "meta" and "genomics," which mean "the study of the microbial genome,". In it, microbiota from different environmental samples is identified, and characterized. This technique helps us to explore the sequences of the culturable and unculturable microbial community that can be explored for the betterment [30–32]. The primary processes in metagenomics involve extracting the entire genome, fragmenting the collected DNA with restriction enzymes, and putting it into an appropriate expression vector. It has been established that the gene codes for the proteins and/or enzymes participating in the biosurfactant synthesis pathway [33] are generally aggregated in the region of the chromosome, and the gene cluster is conjured up of between 3000 and 7000 base pairs. There are metagenomic libraries available for biosurfactants, including function-based techniques like Substrate-Induced Gene Expression-(SIGEX) and High-Throughput-(HTP) screening [34–36]. It is reported that the genomic research is integrated with bioinformatics such as phylogenetic analysis, taxonomic profiling, molecular phylogeny, functional characterization of metagenomes, enzyme research, and system biology studies, including genetic engineering utilizing CRISPR (Clustered Regularly Interspaced Short Palindromic Repeats) or Transcription Activator-like Effector Nucleases (TALENs) [37] that enhances structural and functional diversity knowledge about BS. There is a system database known as "BioSurfDB", that has compiled 58 pathways, 96 biosurfactants, 1077 microorganisms, 3736 genes, and 3430 proteins [38]. Genetic engineering techniques, however, only create small or single-gene alterations. To synthesize a good quantity of biosurfactants, more experiments that are based on optimizations are required. New regulatory aspects should be explored, and biosurfactant-producing genes should be transferred to indigenous microorganisms by exploring the importance of CRISPR-based methods [31], that are residing in various habitats which need more attention in the present era.

6. Biosurfactants as biopesticide and their mechanisms

Pests including insects, flies, rats, nematodes, bacterial and fungal infections, and unfavorable plants influence the health, cultivation, and crop yields. Surfactants have considerable structural features because they are employed as adjuvants in the preparation of pesticides to strengthen their efficacy [27]. They are widely used to prevent the growth of the fungus *Aspergillus* species, which causes aflatoxin contamination of crops [39] and to exterminate weed species that reduce land production [40]. On the other hand, its frequent use affects the health of the soil, human, and environment in a drastic manner. The reason for such a cause is its perilous, persistent, recalcitrant, and extended half-life properties. As a result, it is a challenge for researchers and scientists to find an eco-friendly approach to control this harm and pressure on the environment. Green chemistry, which may create highly valuable products and advance green technology, can overcome such problems. By suppressing the proliferation of pests or solubilizing insoluble compounds, the secondary metabolites generated by microorganisms are well recognized for their beneficial aspects on soil and plant health [25, 41, 42]. However, obtaining and using these surfactants from microbes can be a sustainable strategy for maintaining plant, soil, environmental, and human health. There are several biosurfactants derived from microbes that show antimicrobial activity against deleterious microbes that affect plants. As a result, they are proving themselves as promising biocontrol molecules. The various mechanism that highlights the efficiency of BS to be use as biopesticide are as follows:

6.1 Impact on phytopathogens

Antagonistic effects of BS are seen on various plant pathogens such as *Trichophyton rubrum, Colletotrichum gloeosporioides*, *Corynespora cassiicola, Fusarium verticilliodes, Fusarium oxysporum* f. sp. pisi, and [43], *Aspergillus flavus*, *Penicillium roqueforti*, and *Colletotrichum demiatium* [44, 45], *Fusarium graminearum*, *Botrytis cinerea*, and *Podosphaera fusca* [46–48]. *Colletotrichum capsici* [49], *Macrophomina phaseolina* [50], *Fusarium verticillioides*, *Penicillium* sp., and *Aspergillus* sp. [51].

BS inhibits the growth of these pathogens by formation of an antibiofilm layer on various surfaces, which lowers their capacity to connect, structural failure to the intercellular net and conidiophores, and deferred or lacking sporulation, which leads to a reduction in biomass production [52]. BS also causes cellular membrane destabilization, that disrupts the feeding cycle, and eventually, cell rupture leading to lysis [53, 54]. Biosurfactants' fatty acid components modulate the protein and lipid composition of phytopathogens' cell membranes, altering the osmolarity of the cell and its cell wall structure, and making the pathogens highly receptive to them [55–57]. According to Toral [58] and Hansen [59], BS causes morphological changes in the fungal structure known as hyphae, including swelling of the hyphae, altered mitochondrial organization, lower intracellular pH, esterase activity, and decreased hydrophobicity. BS exhibits more activity against spores [51] than it does against mycelia; this discrepancy in activity may be caused by the differing compositions of mycelia and spores' cell walls [60]. An alternate method to lessen infestations with severe phytopathogens is to strengthen plants' innate immunity that can be seen by BS implementation [61].

6.2 Impact of insects

It represses the short-term attachment of hydroid larvae (*Dynarnena pumila, Obelia loveni* and *Drosophila melanogaster*) and counteracts an attachment and contractility [62–65]. In aphids it damages their cuticle membranes.

7. Biosurfactants and their mechanisms involved in remediating synthetic pesticides

During green revolution, farmers used chemicals such as fertilizers and pesticides (insecticides, herbicides, rodenticides, and fungicides) to satisfy the demand for food. It was noticed only 1% of sprayed pesticides kill their intended target species; the remaining 99 interacts with the soil and produce more complex metabolites that damage the ecosystem and notably the soil quality and human health [66]. Pesticides strongly adsorb organic matter from the soil, preventing it from desorbing. Most pesticides are non-polar chemicals with hydrophobic characteristics that make them water insoluble [67]. The recalcitrant nature of pesticides contributes to a variety of central nervous system problems. In addition to this, pesticides have a huge mutagenic and carcinogenic potential which can result in conditions affecting fertility, skin, and eye problems. It is documented, every year 355,000 persons die from accidental poisonings, which are linked to excessive exposure and inappropriate exposure to dangerous substances [68].

The biodegradation of pesticides by microorganisms through the production of their metabolites has been investigated for the past two decades. Macro and microorganisms use it as it is readily available in ecosystems [18, 69] where many times they consume it as a source of food. When these microbes utilize it, they produce significant metabolites. One characteristic that distinguishes biosurfactants is hydrophilic-lipophilic balance (HLB), which regulates the hydrophilic and hydrophobic constituents balance in surface-active substances [70, 71]. BS is one such prominent metabolite, when present at amounts above CMC, biosurfactant-produced micelles may enhance the solubility and bioavailability of hydrophobic pesticide compounds by lowering interfacial tension and surface tension [72, 73]. Biosurfactant activities rely on the concentration of surface-active molecules until the critical micelle concentration (CMC) is attained. To lower surface tension, efficient biosurfactants have a low CMC and necessitate less biosurfactants [70, 74]. Desorption from soil particles lowers surface tension, facilitating the process of deterioration [75]. One of the most prevalent mechanisms that contribute greatly toward the bioremediation of pesticides by biosurfactants contains counter-ion binding, electrostatic interactions, ion exchange, and precipitation-dissolution [76, 77]. As a result, the soil is made productive, pollution-free, and fit for agricultural cultivation [78, 79]. A natural, budget-friendly, and environmentally sustainable method of degrading pesticides and other xenobiotics on-site is only possible through microbial biosurfactant-based remediation [18]. Although most research highlights where a single bacterium is used in bioremediation (particularly culturable ones), while only a few reports highlight the use of microbial consortia [72, 80]. The capabilities of biosurfactants to remediate environmental pollutants consequently, the burgeoning environmental safety campaign toward greener technologies are the only concerns of the times [25, 26, 78, 81, 82] that can help goal of European Commission to reduce 50% pesticide pollution by 2030.

8. Conclusion and prospects

To assure the prerequisite of food for a growing population is becoming a big challenge at a global level. Regular use of synthetic chemicals is not always recommendable as they have detrimental impacts on the health of plants, soil, and humans. New strategies to enhance agriculture and food production have appeared in the last decades. Microbes and their metabolites always occupy a special place in all related domains of Biotechnology and allied Sciences. Their stability and diversity always accomplish the goals of sustainability. They also play important role in change in climate action. Among all metabolites reported till date, the multifunctional properties of biosurfactants are making this molecule popular in various domains. Therefore, it is advocated that a cutting-edge notion for evaluating a new, safer, and healthier agricultural and environmental model should be explored more and more.

In the chapter a comprehensive information about various methods employed to identify the biosurfactant ability of the microbe, methods to characterize the chemical structure is discussed. The various mechanisms such as competition, parasitism, antibiosis, induced systemic resistance, and hypovirulence by which BS controls the growth of pests are mentioned. Importance of BS as adjuvants in pesticide formations and their mechanisms in pesticide remediation is described. It is recommended that more attention should be given to this molecule and its mechanisms can be explored for other harmful pests which are not documented to date. Its high stability to varied environmental conditions can be used year-round which highlights its additional advantage in the era where a change in climate is also an immediate concern.

Acknowledgements

SG would like to thank Dr. Vishwanath Karad MIT-World Peace University.

Conflict of interest

The author declares no conflict of interest. The author has contributed to the chapter and agreed for its publication.

Author details

Shikha Gaikwad
School of Biosciences and Technology, Dr. Vishwanath Karad MIT World Peace University, Pune, Maharashtra, India

*Address all correspondence to: drshikhagaikwad@gmail.com

References

[1] Sharma A, Kumar V, Shahzad B, Tanveer M, Sidhu GPS, Handa N, et al. Worldwide pesticide usage and its impacts on ecosystem. SN Applied Sciences. 2019;**1**:1-16. DOI: 10.1007/s42452-019-1485-1

[2] Velásquez AC, Castroverde CDM, He SY. Plant–pathogen warfare under changing climate conditions. Current Biology. 2018;**28**(10):R619-R634. DOI: 10.1016/j.cub.2018.03.054

[3] Khatoon Z, Huang S, Rafique M, Fakhar A, Kamran MA, Santoyo G. Unlocking the potential of plant growth-promoting rhizobacteria on soil health and the sustainability of agricultural systems. Journal of Environmental Management. 2020;**273**:111118-111118. DOI: 10.1016/j.jenvman.2020.111118

[4] Ma CS, Zhang W, Peng Y, Zhao F, Chang XQ, Xing K, et al. Climate warming promotes pesticide resistance through expanding overwintering range of a global pest. Nature Communications. 2021;**12**(1):1-10. DOI: 10.1038/s41467-021-25505-7

[5] Gil S, Keeley JDB, Jansen K. The Use of Pesticides in Developing Countries and their Impact on Health and the Right to Food. Brussels: Policy Department for External Relations Directorate General for External Policies of the Union; 2021. DOI: 10.2861/28995

[6] Boedeker W, Watts M, Clausing P, Marquez E. The global distribution of acute unintentional pesticide poisoning: Estimations based on a systematic review. BMC Public Health. 2020;**20**:1-19. DOI: 10.1186/s12889-020-09939-0

[7] Sabarwal A, Kumar K, Singh RP. Hazardous effects of chemical pesticides on human health–Cancer and other associated disorders. Environmental Toxicolology and Pharmacology. 2018;**63**:103-114. DOI: 10.1016/j.etap.2018.08.018

[8] Rani L, Thapa K, Kanojia N, Sharma N, Singh S, Grewal AS, et al. An extensive review on the consequences of chemical pesticides on human health and environment. Journal of Cleaner Production. 2021;**283**:124657. DOI: 10.1016/j. jclepro.2020.124657

[9] Nicolopoulou-Stamati P, Maipas S, Kotampasi C, Stamatis P, Hens L. Chemical pesticides and human health: The urgent need for a new concept in agriculture. Frontiers in Public Health. 2016;**4**:148. DOI: 10.3389/fpubh.2016.00148

[10] Akhtar FZ, Archana KM, Krishnaswamy VG, Rajagopal R. Remediation of heavy metals (Cr, Zn) using physical, chemical and biological methods: A novel approach. SN Applied Sciences. 2020;**2**:1-14. DOI: 10.1007/s42452-019- 1918-x

[11] Castro MJ, Ojeda C, Cirelli AF. In: Lichtfouse E, Schwarzbauer J, Robert D, editors. Advances in Surfactants in Agriculture in Green Materials for Energy, Products and Depollution. Switzerland: Springer; 2013. pp. 287-334. DOI: 10.1007/ 978-94-007-6836-9_7

[12] Ahuja K, Singh S. Biosurfactants market size by products. Glob Mark Insight. 2020. Available from: https://www.gminsights.com/industry-analysis/biosurfactants-market-report [Accessed Apr 5, 2021]

[13] Soberón-Chávez G, Lépine F, Déziel E. Production of rhamnolipids by *Pseudomonas aeruginosa*. Applied

Microbiology and Biotechnology. 2005;**68**:718-725. DOI: 10.1007/s00253-005-0150-3

[14] Moya Ramírez I, Tsaousi K, Rudden M, Marchant R, Jurado Alameda E, García Román M, et al. Rhamnolipid and surfactin production from olive oil mill waste as sole carbon source. Bioresource and Technology. 2015;**98**:231-236. DOI: 10.1016/j.biortech.2015.09.012

[15] Drakontis CE, Amin S. Biosurfactants: Formulations, properties, and applications. Current Opinion in Colloid & Interface Science. 2020;**48**:77-99. DOI: 10.1016/j.cocis.2020.03.013

[16] Sarubbo LA, Maria da Gloria CS, Durval IJB, Bezerra KGO, Ribeiro BG, Silva IA, et al. Biosurfactants: Production, properties, applications, trends, and general perspectives. Biochemical Engineering Journal. 2022;**181**:108377. DOI: 10.1016/j.bej. 2022.108377

[17] Banat IM, Satpute SK, Cameotra SS, Patil R, Nyayanit NV. Cost effective technologies and renewable substrates for biosurfactants' production. Frontiers in Microbiology. 2014;**5**:697. DOI: 10.3389/fmicb.2014.00697

[18] Lamilla C, Schalchli H, Briceño G, Leiva B, Donoso-Piñol P, Barrientos L, et al. A pesticide biopurification system: A source of biosurfactant producing bacteria with environmental biotechnology applications. Agronomy. 2021;**11**:624. DOI: 10.3390/agronomy11040624

[19] Naughton PJ, Marchant R, Naughton V, Banat IM. Microbial biosurfactants: Current trends and applications in agricultural and biomedical industries. Journal of Applied Microbiology. 2019;**127**:12-28. DOI: 10.1111/jam.14243

[20] Jahan R, Bodratti AM, Tsianou M, Alexandridis P. Biosurfactants, natural alternatives to synthetic surfactants: Physicochemical properties and applications. Advances in Colloid Interface Science. 2020;**275**:102061. DOI: 10.1016/j.cis.2019.102061

[21] Crouzet J, Arguelles-Arias A, Dhondt-Cordelier S, Cordelier S, Pršic J, Hoff G, et al. Biosurfactants in plant protection against diseases: Rhamnolipids and lipopeptides case study. Frontiers in Bioengineering and Biotechnology. 2020;**8**:1014. DOI: 10.3389/fbioe.2020.01014

[22] Adetunji C O, Jeevanandam J, Selvam J D, Inobeme A, Thangadurai D, Michael O S, et al. Production and commercialization of biosurfactants for plant pest management, in Applications of biosurfactant in agriculture. Inamuddin and C. O. Adetunji (Massachusetts: Academic Press), 2022; 69-92. doi: 10.1016/B978-0-12-822921-7.00003-9

[23] Ceresa C, Fracchia L, Fedeli E, Porta C, Banat IM. Recent advances in biomedical, therapeutic and pharmaceutical applications of microbial surfactants. Pharmaceutics. 2021;**13**(4):466. DOI: 10.3390/pharmaceutics13040466

[24] Kumar A, Singh SK, Kant C, Verma H, Kumar D, Singh PP, et al. Microbial biosurfactant: A new frontier for sustainable agriculture and pharmaceutical industries. Antioxidants. 2021;**10**:1472. DOI: 10.3390/antiox10091472

[25] Karamchandani BM, Maurya PA, Dalvi SG, Waghmode S, Sharma D, Rahman PKSM, et al. Synergistic activity of rhamnolipid biosurfactant and nanoparticles synthesized using fungal origin chitosan against phytopathogens.

Frontiers of Bioengineering and Biotechnology. 2022;**10**:917105. DOI: 10.3389/fbioe.2022.917105

[26] Karamchandani BM, Chakraborty S, Dalvi SG, Satpute SK. Chitosan and its derivatives: Promising biomaterial in averting fungal diseases of sugarcane and other crops. Journal of Basic Microbiology. 2022;**62**:533-554. DOI: 10.1002/jobm.202100613

[27] Adetunji CO, Jeevanandam J, Inobeme A, Olaniyan OT, Anani OA, Thangadurai D, et al. Application of biosurfactant for the production of adjuvant and their synergetic effects when combined with different agro-pesticides. Green Sustainable Process for Chemical and Environmental Engineering and Science. 2021:255-277. DOI: 10.1016/b978-0-12-823380-1.00006-x

[28] Sánchez C. A review of the role of biosurfactants in the biodegradation of hydrophobic organo pollutants: Production, mode of action, biosynthesis and applications. World Journal of Microbiology and Biotechnology. 2022;**38**(11):216-224. DOI: 10.1007/s11274-022-03401-6

[29] Hamzah AF, Al-Mossawy MI, Al-Tamimi WH, et al. Enhancing the spontaneous imbibition process using biosurfactants produced from bacteria isolated from Al-Rafidiya oil field for improved oil recovery. Journal of Petrol Exploration and Production Technology. 2020;**10**:3767-3777. DOI: doi.org/10.1007/s13202-020-00874-9

[30] Malla MA, Dubey A, Kumar A, Yadav S, Hashem A, Abd Allah EF. Exploring the human microbiome: The potential future role of next generation sequencing in disease diagnosis and treatment. Frontiers in Immunology. 2019;**9**:2868. DOI: 10.3389/fimmu.2018.02868

[31] Kumar A, Dubey A. Rhizosphere microbiome: Engineering bacterial competitiveness for enhancing crop production. Journal of Advanced Research. 2020;**24**:337-352. DOI: 10.1016/j.jare.2020.04.014

[32] Raj A, Kumar A, Dames JF. Tapping the role of microbial biosurfactants in pesticide remediation: An eco-friendly approach for environmental sustainability. Frontiers in Microbiology. 2021;**12**:791723. DOI: 10.3389/fmicb.2021.791723

[33] Araújo WJ, Oliveira JS, Araújo SCS, Minnicelli CF, Silva-Portela RCB, da Fonseca MMB, et al. Microbial culture in minimal medium with oil favors enrichment of biosurfactant producing genes. Frontiers in Bioengineering and Biotechnology. 2020;**8**:962. DOI: 10.3389/fbioe.2020.00962

[34] Datta S, Rajnish KN, Samuel MS, Pugazlendhi A, Selvarajan E. Metagenomic applications in microbial diversity, bioremediation, pollution monitoring, enzyme and drug discovery. A review. Environmental Chemistry Letters. 2020;**18**:1229-1241. DOI: 10.1007/s10311-020-01010-z

[35] Femina Carolin C, Kumar PS, Joshiba GJ, Ramamurthy R, Varjani SJ. Bioremediation of 2,4-diaminotoluene in aqueous solution enhanced by lipopeptide biosurfactant production from bacterial strains. Journal of Environmental Engineering. 2020;**146**:04020069. DOI: 10.1061/(asce)ee.1943-7870.0001740

[36] Femina Carolin C, Kumar PS, Joshiba GJ, Madhesh P, Ramamurthy R. Sustainable strategy for the enhancement of hazardous aromatic amine degradation using lipopeptide biosurfactant isolated from *Brevibacterium casei*. Journal of Hazardous Material. 2021;**408**:124943. DOI: 10.1016/j.jhazmat.2020.124943

[37] Singh S, Singh H, Rout B, Tripathi RBM, Chopra C, Chopra RS. The new science of metagenomics: Revealing the secrets of microbial physiology. In: Chopra RS, Chopra C, Sharma NR, editors. Metagenomics: Techniques, Applications, Challenges and Opportunities. Singapore: Springer; 2020. pp. 3-22. DOI: 10.1007/978-981-15-6529-8_1

[38] Kumari P, Kumar Y. Bioinformatics and computational tools in bioremediation and biodegradation of environmental pollutants. In: Saxena G, Kumar V, Shah M, editors. Bioremediation for Environmental Sustainability. Amsterdam: Elsevier; 2021. pp. 421-444

[39] Rodriguez SB, Mahoney NE. Inhibition of aflatoxin production by surfactants. Applied Environmental Microbiology. 2020;**60**:106-110

[40] Boyette CD, Walker HL, Abbas HK. Biological control of kudzu (*Pueraria lobata*); with an isolate of *Myrothecium verrucaria*. Biocontrol Science and Technology. 2002;**12**:75-78

[41] Adnan M, Alshammari E, Ashraf SA, Patel K, Lad K, Patel M. Physiological and molecular characterization of biosurfactant producing endophytic fungi *Xylaria regalis* from the cones of *Thuja plicata* as a potent plant growth promoter with its potential application. Biomed Research International. 2018;**2018**:1-11. DOI: 10.1155/2018/7362148

[42] Chopra A, Vandana UK, Rahi P, Satpute S, Mazumder PB. Plant growth promoting potential of *Brevibacterium sediminis* A6 isolated from the tea rhizosphere of Assam, India. Biocatalysis and Agricultural Biotechnology. 2020;**27**:101610. DOI: 10.1016/j.bcab.2020.101610

[43] Suparna S, Narayan BS, Arijit B, Suresh D. Production, characterization, and antifungal activity of a biosurfactant produced by *Rhodotorula babjevae* YS3. Microbial Cell Factories. 2017;**16**:95

[44] Yu GY, Sinclair JB, Hartman GL, Bertagnolli BL. Production of Iturin A by *Bacillus amyloliquefaciens* suppressing *Rhizoctonia solani*. Soil Biology and Biochemistry. 2002;**34**(7):955-963

[45] Velmurugan N, Choi MS, Han SS, Lee YS. Evaluation of antagonistic activities of *Bacillus subtilis* and *Bacillus licheniformis* against wood-staining fungi: *In vitro* and *in vivo* experiments. Journal of Microbiology. 2009;**47**:385-392

[46] Wang J, Yao J, Liu J, Chen H, Yao J. Characterization of *Fusarium graminearum* inhibitory lipopeptide from *Bacillus subtilis* IB. Applied Microbiology and Biotechnology. 2007;**76**:889-894

[47] Romero D, Vicente A d, Rakotoaly RH, et al. Iturin and fengycin families of lipopeptides are key factors in antagonism of *Bacillus subtilis* toward *Podosphaera fusca*. Molecular Plant-Microbe Interactions. 2007;**2007**(20):430-440

[48] Pang AZ, Wang F, Baocai X, Hesham R. Lipopeptide biosurfactants from *Bacillus spp.*: Types, production, biological activities, and applications in food nawazish. El-Seedi Journal of Food Quality. 2022;**3930112**:19

[49] Jiumoni L, Debahuti G, Suresh D, Giasuddin A. Novel approaches for application of biosurfactant produced by *Pseudomonas aeruginosa* for biocontrol of *Colletotrichum capsici* responsible for anthracnose disease in chilli. European Journal of Plant Pathology. 2017;**150**(1):57-71. DOI: 10.1007/s10658-017-1252-3

[50] Ekta K, Naveen AK. Biosurfactant based formulation of *Pseudomonas*

guariconensis LE3 with multifarious plant growth promoting traits controls charcoal rot disease in *Helianthus annus*. World Journal of Microbiology and Biotechnology. 2021;**37**:55. DOI: 10.1007/s11274-021-03015-4

[51] Borah SN, Goswami D, Sarma HK, Cameotra SS, Deka S. Rhamnolipid biosurfactant against *Fusarium verticillioides* to control stalk and ear rot disease of maize. Frontiers in Microbiology. 2016;**7**:1505. DOI: 10.3389/fmicb.2016.01505

[52] Mirela MG, Matei S, Matei A, Draghici E. Antifungal activity of a biosurfactant-producing lactic acid bacteria strain. The EuroBiotech Journal. 2017;**212**:1-3

[53] Hingley ST, Hastie AT, Kueppers F. Effect of ciliostatic factors from Pseudomonas aeruginosa on rabbit respiratory cilia. Infection and Immunity. 1986;**51**:254-258

[54] Banat IM, Franzetti A, Gandolfi I, Bestetti G, Martinotti MG, Fracchia L, et al. Microbial biosurfactants production, applications and future potential. Applied Microbiology and Biotechnology. 2010;**87**:427-444

[55] Desai JD, Banat IM. Microbial production of surfactants and their commercial potential. Microbiology and Molecular Biology Reviews. 1997;**61**:47-56

[56] Mawgoud AM, Lépine F, Déziel E. Rhamnolipids: Diversity of structures, microbial origins and roles. Applied Microbiology and Biotechnology. 2010;**86**:1323-1336. DOI: 10.1007/s00253-010-2498-2

[57] Gomaa EZ. Antimicrobial activity of a biosurfactant produced by *Bacillus licheniformis strain* M104 grown on whey. Brazilian Archives of Biology and Technology. 2013;**56**:259-268

[58] Toral L, Rodr'ıguez M, Bejar V, Sampedro I. Anti-fungal activity of lipopeptides from *Bacillus* XT1 CECT 8661 against *Botrytis cinerea*. Frontiers in Microbiology. 2018;**2018**(9):1315-1412

[59] Hansen MC, Olsson RS, Sorensen J. Confocal imaging of living fungal hyphae challenged with the fungal antagonist viscosinamide. Mycologia. 2000;**92**:216-221

[60] Kaur S, Dhillon GS, Brar SK, Vallad GE, Chand R, Chauhan VB. Emerging phytopathogen *Macrophomina phaseolina*: Biology, economic importance and current diagnostic trends. Critical Reviews in Microbiology. 2012;**38**(2):136-151

[61] Vatsa P, Sanchez L, Clément C, Baillieul F, Dorey S. Rhamnolipid biosurfactants as new players in animal and plant defense against microbes. International Journal of Molecular Sciences. 2010;**11**:5095-5108

[62] Railkin AI. Behavioral and physiological reactions of hydroids and Bivalves to some antifouling substances. Zoologicheskii Zhural. 1994;**73**(7-8):22-30

[63] Gronwald JW, Plaisance KL, Ide DA, Wyse DL. Assessment of *Pseudomonas syringae pv. tagetis* as a biocontrol agent for Canada thistle. Weed Science. 2002;**50**:397-404

[64] Krishnayya PV, Grewal PS. Effect of neem and selected fungicides on viability and virulence of the entomopathogenic nematode *steinernema feltiae*. Biocontrol Science and Technology. 20o2;**12**:259-252

[65] Jazzar C, Hammad EA. The efficacy of enhanced aqueous extracts of melia azedarach leaves and fruits integrated with the *Camptotylus reuteri* releases against the sweet potato whitefly

nymphs. Bulletin of Insectology. 2003;**56**:269-275

[66] Rawat G, Dhasmana A, Kumar V. Biosurfactants: The next generation biomolecules for diverse applications. Environmental Sustainability. 2020;**3**:353-369. DOI: 10.1007/s42398-020-00128-8

[67] Kalyani N, Goel S, Jaiswal S. On-site sensing of pesticides using point-of-care biosensors: A review. Environmental Chemistry Letters. 2021;**19**:345-354. DOI: 10.1007/s10311-020-01070-1

[68] Nayak S, Sahoo A, Kolanthasamy E, Rao K. Role of pesticide application in environmental degradation and its remediation strategies. In: Kumar V, Singh J, Kumar P, editors. Environmental Degradation: Causes and Remediation Strategies. Vol. 1. Haridwar: Agriculture and Environmental Science Academy; 2020. p. 36. DOI: 10.26832/aesa-2020-edcrs-03

[69] García-Reyes S, Yáñez-Ocampo G, Wong-Villarreal A, Rajaretinam RK, Thavasimuthu C, Patiño R, et al. Partial characterization of a biosurfactant extracted from *Pseudomonas* sp. B0406 that enhances the solubility of pesticides. Environmental Technology. 2018;**39**:2622-2631

[70] Pacwa-Płociniczak M, Płaza GA, Piotrowska-Seget Z, Cameotra SS. Environmental applications of biosurfactants: Recent advances. International Journal of Molecular Science. 2011;**12**:633-654. DOI: 10.3390/ijms12010633

[71] Fenibo EO, Ijoma GN, Selvarajan R, Chikere CB. Microbial surfactants: the next generation multifunctional biomolecules for applications in the petroleum industry and its associated environmental remediation. Microorganisms. 2019;**7**:581. DOI: 10.3390/microorganisms7110581

[72] Rasheed T, Shafi S, Bilal M, Hussain T, Sher F, Rizwan K. Surfactants-based remediation as an effective approach for removal of environmental pollutants—A review. Journal of Molecular Liquids. 2020;**318**:113960. DOI: 10.1016/j.molliq.2020.113960

[73] Bami MS, Estabragh MAR, Ohadi M, Banat IM, Dehghannoudeh G. Biosurfactants aided bioremediation mechanisms: A mini-review. Soil and Sediment Containation: An International Journal. 2022;**31**:801-817. DOI: 10.1080/15320383.2021.2016603

[74] Bhatt P, Verma A, Gangola S, Bhandari G, Chen S. Microbial glycoconjugates in organic pollutant bioremediation: Recent advances and applications. Microbial Cell Factories. 2021;**20**:1-18. DOI: 10.1186/s12934-021-01556-9

[75] Twigg M, Tripathi L, Zompra K, Salek K, Irorere V, Gutierrez T, et al. Surfactants from the sea: Rhamnolipid production by marine bacteria. Access Microbiology. 2019;**1**:192. DOI: 10.1099/acmi.ac2019.po0066

[76] Xu X, Liu W, Tian S, Wang W, Qi Q, Jiang P, et al. Petroleum hydrocarbon-degrading bacteria for the remediation of oil pollution under aerobic conditions: A perspective analysis. Frontiers in Microbiology. 2018;**9**:2885. DOI: 10.3389/fmicb.2018.02885

[77] Pardhi DS, Panchal RR, Raval VH, Joshi RG, Poczai P, Almalki WH, et al. Microbial surfactants: A journey from fundamentals to recent advances. Frontiers in Microbiology. 2022;**13**:982603. DOI: 10.3389/fmicb.2022.982603

[78] Jimoh AA, Lin J. Enhancement of *Paenibacillus* sp. D9 lipopeptide

biosurfactant production through the optimization of medium composition and its application for biodegradation of hydrophobic pollutants. Applied Biochemistry and Biotechnology. 2019;**187**:724-743. DOI: 10.1007/s12010-018-2847-7

[79] Yesankar PJ, Pal M, Patil A, Qureshi A. Microbial exopolymeric substances and biosurfactants as 'bioavailability enhancers' for polycyclic aromatic hydrocarbons biodegradation. International Journal of Environmental Sciences and Technology. 2023;**20**:5823-5844. DOI: 10.1007/s13762-022-04068-0

[80] Madamwar D, Tsuda M, Solyanikova I, Chen S, Mishra S, Lin Z, et al. Recent advanced technologies for the characterization of xenobiotic-degrading microorganisms and microbial communities. Frontiers Bioengineering and Biotechnology. 2021;**9**:632059. DOI: 10.3389/fbioe.2021.632059

[81] Odukkathil G, Vasudevan N. Enhanced biodegradation of endosulfan and its major metabolite endosulfate by a biosurfactant producing bacterium. Journal of Environmental Sciences and Health. 2013;**48**:462-469

[82] Patowary K, Patowary R, Kalita MC, Deka S. Characterization of biosurfactant produced during degradation of hydrocarbons using crude oil as sole source of carbon. Frontiers in Microbiology. 2017;**8**:279. DOI: 10.3389/fmicb.2017.00279

Chapter 5

Perspective Chapter: Genomics, Proteomics, and System Biology of Insecticides Resistance in Insects

Rabbiah Manzoor Malik, Sahar Fazal, Narjis Khatoon, Muneeba Ishtiaq, Saima Batool and Syed Tauqeer Abbas

Abstract

Insecticide resistance is an inherited change in pest population exposure to a specific insecticide or group of insecticides. Overuse, misuse, and high interbreeding rates have led to insecticide resistance. Genomic technologies reveal mechanisms of resistance, including decreased target-site sensitivity and increased detoxification. Genomic projects have cloned and identified targeted genes in Drosophila melanogaster and studied resistance-associated mutations in various pest insects. Advancements in genome sequencing and annotation techniques have explored complex multigene enzyme systems, such as glutathione-S-transferases, esterases, and cytochrome P450, which facilitate insecticide resistance. Identifying specific genes involved in resistance and targeted genes is essential for developing new insecticides and strategies to control pests. Insects with resistance metabolize insecticidal compounds faster due to increased catalytic rate and gene amplification. So, system biology plays a very important role in the insect resistance against insecticides and different chemicals such as DDT and permethrin. From system biology, not only the identification of genes was done, but also the protein-protein interactions were found out, which were responsible in the insect resistance.

Keywords: insecticides resistance, system biology of insecticide resistance, p450 and insecticides resistance, genomics of insecticides, proteomics of insecticides

1. Introduction

The most diverse group of animals on Earth is insect, which performs a lot of significant roles. The important functions of the insects are that they act as decomposers of the dead organisms, they are the necessary components of the ecosystems, they are helpful in the pollination of the plants, they also spread the seeds, they are a good source of proteins for the livestock, and they provide us with a variety of products such as dye, silk, wax, and honey [1–4]. Though insects are playing an important role in the lives of humans and livestock, they also prove to be very harmful for the environment and humans [5]. The negative effects of insects on the environment and humans are the major threats for the ecosystem. The most highlighted negative impact of insects is on agriculture. Some insects such as locusts, caterpillars, and grasshoppers act as pests for crops because they eat the fruits, seed, and leaves of the crops. Some insects affect the development and the growth of the crops and make the plants vulnerable to the diseases; these insects include thrips, weevils, and aphids. Some insects such as locusts damage the crops, which results in famine situations. The pests have negative impacts on the agricultural crops as they spread and carry different diseases of the plants [5, 6]. These harmful impacts of the insects on the crops are resulting in the food shortage in the different parts of the world because the population of the world is increasing rapidly. The damages caused by the insects are also increasing with the increase in climate change [7]. Insects are the main source of causing the infectious diseases among the humans and livestock; for example, mosquitoes are responsible for causing malaria and dengue. Other insects involved in causing infectious diseases are kissing bugs, head lice, body lice, tsetse flies, and so on [8]. Keeping in view these all negative impacts of insects on the humans and environment, there is a need to control the insects.

2. Control of insects and insecticide resistance

Insecticides are used for the control of insects (such as termites, cockroaches, lice, and mosquitoes) in public health, industries, households, and agriculture. Firstly, the DDT was used to control the insects, but the insects got resistant to DDT, and it was reported in the houseflies in 1947. The new and most widely used insecticides are carbamates, organophosphates, formamidines, neonicotinoids, and pyrethroids. These insecticides were very effective against the insects in the beginning, but with the passage of time, insects became resistant to these insecticides [9, 10]. An inherited change in the exposure of a population of a pest to a certain insecticide or a group of insecticides is called insecticide resistance. The insects that cannot be controlled by the repeated use of a particular insecticide are said to be resistant to that insecticide [11]. The insects are getting resistant to the insecticides quickly because the insecticides are overused or misused, the insects' population is greater in size, and they interbreed at a very high rate [12].

3. Genomics and its significance in the field of biological sciences

An interdisciplinary field of biology that focuses on the function, structure, mapping, genome editing, and evolution is called genomics. The complete set of the DNA of an organism is called the genome. The aim of genomics is to study all the collective quantification and characterization of the genes of an organism and their impact on an organism [13]. The field of the biological sciences has become advanced with the help of the genomics as it involves the analysis and the sequencing of the genomes by using the next-generation sequencing and computational tools to analyze and assemble the structure and functions of the genomes [14]. Genomics has brought a revolution in the field of biological sciences such as systems biology, discovery-based research, biotechnology, medical diagnosis, personalized medicines, identifying therapeutic targets, forensics, biology systematics, and finding the evolutionary histories of the organisms. Intra-genomic studies are also involved in genomics such as pleiotropy, epistasis, heterosis, and the interactions between the alleles and loci within the genome [15].

4. Genomics and insecticide resistance

The technologies of genomics are showing different mechanisms of insecticide resistance, which involves decreased target-site sensitivity and increased detoxification [16]. Some possibly important concerns related to the quick insecticide resistance among insects with the evolutionary time are also revealed by the genome projects. Evolutionary biologists are being provided with contemporary and ideal model systems to study the evolution of the resistance among insects for the insecticides [17, 18]. The use of the tools of molecular biology to eliminate the mechanisms of insecticide resistance is of great interest. In 1990s, traditional techniques of molecular biology were used to investigate a few cases of insecticide resistance at a molecular level. The cases that involved the known genes could easily be cloned with the heterologous PCR were manageable. From the early studies, three mechanisms of the insecticides resistance were reported; one mechanism involved target-site insecticide resistance, and the other two mechanisms involved the increased detoxification of the insecticides. In culex mosquitoes [19] and aphids [20], the resistance against carbamate and organophosphate has been reported, and it is an example of the mechanism of detoxification.

Another example of detoxification is in the two species of flies in which the degradation of insecticides takes place. In specific carboxylesterases, the structural mutations had arisen that used to convert them into inefficient but physiologically sufficient organophosphate hydrolases [21]. The third mechanism of detoxification is the mutation of the target molecule in such a way that the target molecule becomes insensitive to the insecticides. The target molecules that become mutant are: for cyclodienes, the γ-aminobutyric acid (GABA) receptors become mutant; for organophosphates, the acetylcholinesterase becomes mutant; and for synthetic pyrethroids and dichlorodiphenyltrichloroethane (DDT), the voltage-gated sodium channels are becoming mutant [22, 23]. These findings were having some remarkable aspects such as the degradation and sequestration mechanisms, the sodium channels becoming insensitive, and repetition of the same amino acid changes in the orthologous proteins among different species, for example, in the acetyl cholinesterases and GABA receptors. The third aspect of these findings was that within a few years of the first use of insecticide, a small amount of the mutant alleles carry the mutations that have spread among the species. These features have shown that insects have very less options to confer the resistance against insecticides [24, 25].

Genomic technologies are able to investigate the previous intractable mechanisms of the resistance. Genomics also discusses the resistance to the proteinaceous biopesticide crystal toxins of *Bacillus thuringiensis* (Bt toxins) and the traditional chemical insecticides [26]. By the help of genomics, some targeted genes of the nervous system of *Drosophila melanogaster* have been cloned and identified and in a wide range of the pest insects, the resistance associated mutations have been studied [27]. Recently, with the advancement in genome sequencing and annotation techniques, genomes of the insects have been sequenced and annotated and the complex multigene enzyme systems such as the glutathione-S-transferases, esterases, and cytochrome P450 that facilitate the insecticides resistance among the insects have been explored [28]. In 2000, the whole genome of *Drosophila melanogaster* was reported, and after that, the partial and complete genomes of different species of insects have started to publish in the biological databases. In the NCBI database, genome sequences of almost 34 species of the orders Hymenoptera, Coleoptra, Diptera, Hemiptera, and Lepidoptera of the insects are available. These species include the most primitive insect human louse and major medical pests such as *Aedes aegyptii* and *Anopheles gambiae* [28].

5. Genes involved in insecticide resistance in insects

Recently, insecticide resistance has become a major concern for the control of many insect pest species. This challenging problem has useful solutions in the genome sequencing, transcriptome analysis, and the global quantization of the gene expression of those genes that are involved in the insecticide resistance. One of the most destructive agricultural pests of the world is *Bactrocera dorsalis* (oriental fruit fly), and it is used as a model to examine the genetic mechanisms of the insecticides resistance. For this species, the molecular data of the genes that were identified by homology was very limited. By using the Ilumina Solexa platform of the next-generation sequencing, the whole transcriptome of *Bactrocera dorsalis* was sequenced and the gene expression in the insecticide resistance was explored [29].

Mosquitoes are the major carriers of pathogens, and they are the source of causing infectious diseases among humans such as dengue and malaria, and the control of mosquitoes is the biggest threat as they are resistant to insecticides. In natural populations, the alternative tools for the control of mosquitoes have been implemented and the mechanism of the resistance was studied. A common mechanism of the resistance is the biodegradation of the insecticides by detoxification enzymes; during this mechanism, the changes in the genome of the mosquitoes have been identified except the individual genotyping of the resistance. Particularly, polymorphisms of the detoxification enzymes and the function of the copy number variations (CNVs) have not been examined at the genomic level though they represent strong markers for metabolic resistance. With the use of next-generation sequencing, the genes and polymorphisms associated with insecticide resistance in mosquitoes have been explored. According to a research, 760 candidate genes were sequenced and identified to be the cause of resistance against deltamethrin in the dengue mosquito (*Aedes aegypti*) [30]. The analysis of the CNVs showed the amplification of 41 genes to be associated with the resistance and in the resistant populations, the cytochrome P450 was over transcribed. More than 30,000 variants were detected in the analysis of the polymorphism. By combing the filtering of allele frequency and the Bayesian 55 nonsynonymous variants that were strongly associated in causing the resistance were identified. Both the polymorphisms and the CNVs within the regions were conserved but differed across the continents, which confirm that the changes in the genome causes the metabolic resistance against insecticides are not universal. The novel DNA markers for insecticide resistance were identified, which open the way for tracing the metabolic changes established by the mosquitoes for resisting the insecticides within and among the populations [31].

Anopheles gambiae is resistant to the four classes of insecticides, that is, the carbamates, pyrethroids, organophosphates, and organochlorines; that is why the control of the malaria is difficult in Africa. The functional validation of the detoxifying enzymes is lacking in *Anopheles gambiae*, but the expression of the detoxifying enzymes increases in resisting the insecticides. In the resistant *Anopheles gambiae*, the three genes Cyp6p3, Cyp6m2, and Gste2 are upregulated; for these findings and to explore the phenotype of the insecticide resistance, the transgenic analysis was performed using the UAS/GAL4 system. The evidence was reported that the resistance against organochlorine and organophosphate in *Anopheles gambiae* explains the overexpression of GSTE2 in a wide tissue profile. Carbamate and pyrethroid resistance is given by the overexpression of Cyp6p3; in the same tissues, pyrethroid resistance is explained by Cyp6m2. According to a research conducted on 757 samples of *Anopheles gambiae*, the mutations in the rdl, ace-1, and kdr gene were detected using sequencing and SNaPshot. In the insecticide resistance in *Anopheles gambiae* populations, the multiple mutations were also detected in the kdrW, ace-1, and A296G rdl alleles [32].

6. Mechanisms of insecticide resistance

Insecticide resistance is primarily caused by changes in the genes of insects. The genes that are involved in the insecticide resistance include those that encode for detoxification enzymes such as cytochrome P450 (CYP) and glutathione S-transferase (GST), which metabolize and detoxify the insecticides. These enzymes can also have mutations that increase their activity, making the insecticides less toxic. Target-site resistance mechanisms are also driven by mutations in the genes encoding the target proteins of the insecticides. On the other hand, insecticides target specific genes in insects to kill them. These genes are responsible for vital processes such as nerve impulse transmission, muscle contraction, and metabolism. For example, many insecticides target the voltage-gated sodium channels in the insects' nervous system, which are necessary for nerve impulse transmission [11, 16].

Other insecticides target enzymes that are involved in the production of energy in the insects, such as the mitochondrial electron transport chain, making it impossible for the insects to survive. Insecticides also target genes that are responsible for the synthesis of chitin, which is an important component of the insects' exoskeleton and necessary for their survival. It is important to note that the mechanisms of resistance and the target of the insecticides are constantly evolving due to the insects' adaptation to the environment and the insecticides. Therefore, the identification of the specific genes involved in resistance and the genes targeted by insecticides is essential for the development of new insecticides and strategies to control the pests [33]. There are different mechanisms of insecticide resistance, which include target-site insecticide resistance, metabolic insecticide resistance, penetration resistance, and behavioral resistance.

7. Penetration resistance

The susceptible insects absorb the toxin more quickly than the resistant insects. When the insects' outer cuticle develops the barriers of the slow absorption of the insecticides in their bodies, the penetration resistance occurs. Due to the penetration resistance, insects are protected from a wide range of the insecticides. Along with the other mechanisms of the insecticides, the penetration resistance takes place, and due to the reduced intensity of the penetration, these mechanisms of resistance dominate among insects [34].

8. Behavioral resistance

The insects that are resistant to insecticides are able to recognize and detect a danger and to avoid the toxin. For various classes of insecticides such as organophosphates, carbamates, organochlorines, and pyrethroids, the behavioral mechanism has been reported [35].

9. Target-site insecticide resistance

The specific binding site of an insecticide is mutated or modified during the resistance of the target site, due to which the target site becomes incompatible for the activation. In most common pests (such as *Myzus persicae*, *Musca domestica*, and *Drosophila melanogaster*), the mutations occur in the target regions, that is, knockdown resistance to pyrethroids, reduced sensitivity of the sodium channels against DDT, and the resistance against spinosad and subunits like nicotinic acetylcholine receptors for the neonicotinoids [24, 25]. Because of these mutations, the binding of the target region with the insecticides becomes impossible, and this leads to a loss of binding affinity. Moreover, the overproduction of the enzymes occurs in the metabolic resistance, which detoxify or break down the insecticides, leading to the resistance of the pests. Some metabolic enzymes such as hydrolases, cytochrome p450 monooxygenase, and glutathione S-transferase play a major role in the evolution of metabolic resistance. In the wild-type AChE gene (ace), the point mutations were found in the resistant *B. dorsalis*. In some species of the insects, the resistance also arises from the novel variants that represent the genetic changes such as the RNA edited product or alternatively spliced RNA [36].

10. Metabolic insecticide resistance

Metabolic insecticide resistance, also known as detoxification-based resistance, is a mechanism by which insects are able to detoxify the toxic compounds present in insecticides through the action of enzymes. This type of resistance is becoming increasingly common and is a significant threat to the control of insect pests. According to recent research, metabolic insecticide resistance has been primarily mediated by the activity of enzymes such as cytochrome P450 monooxygenases (P450s), esterases, and glutathione S-transferases (GSTs). These enzymes are able to detoxify the toxic compounds present in insecticides, rendering them harmless to the insect [10].

11. Proteomics- proteins and compounds involved in developing resistance

One example of metabolic insecticide resistance is found in the cotton bollworm, *Helicoverpa armigera*. Research has shown that this pest is able to detoxify the insecticide deltamethrin through the action of P450 enzymes. Specifically, the study found that the insect had an increased expression of the P450 gene CYP6B8, which was responsible for detoxifying the insecticide. Another example can be found in the red flour beetle, *Tribolium castaneum*. Research has shown that this pest is able to detoxify the insecticide chlorpyrifos through the action of esterases. Specifically, the study found that the insect had an increased activity of the esterase enzyme, which was responsible for detoxifying the insecticide [36–38].

Metabolic insecticide resistance can also be found in the mosquito, *Aedes aegypti*. Research has shown that this pest is able to detoxify the insecticide temephos through the action of GSTs. Specifically, the study found that the insect had an increased activity of the GST enzyme, which was responsible for detoxifying the insecticide. It is important to note that the evolution of resistance in insects is a complex process, influenced by a combination of genetic, biochemical, and environmental factors. To ensure effective control of insect pests, it is crucial to adopt integrated pest management strategies that include the use of insecticides in combination with other control measures such as source reduction, biological control, and the use of alternative treatments such as essential oils. Research shows that metabolic insecticide resistance is a significant problem that is becoming increasingly common. The resistance is primarily mediated by the activity of enzymes such as P450s, esterases, and GSTs, which are able to detoxify the toxic compounds present in insecticides. To effectively control insect pests, it is crucial to adopt integrated pest management strategies that include the use of insecticides in combination with other control measures [30, 31].

The mechanism of the insecticide resistance of some insects is explained here:

11.1 Cockroaches

In more than half a dozen insect pest species, point mutations in the para sodium channel gene have been linked to knockdown resistance (kdr) to pyrethroids insecticides. In this investigation, we found two novel para variants in five strains of German cockroaches with high levels of resistance to kdr. The first intracellular linker, which joins domains I and II, contains the two alterations, which change glutamic acid (E434) to lysine (K434) and cysteine (C764) to arginine (R764), respectively. Closest to domain I is E434K, which is found near the beginning of the linker. C764R is found near the end of the linker (closest to domain II). One of the resistant strains has two further mutations, one from proline (P1880) to leucine (L1888) and another from aspartic acid (D58) to glycine (G58). The four mutations are exclusively seen in the most resistant individuals of a particular strain, and they coexist with the previously discovered leucine to phenylalanine (L993F) kdr mutation in IIS6. These findings imply that these mutations may be in charge of the German cockroach's high levels of knockdown resistance to pyrethroids pesticides [39, 40].

11.2 Head lice

Pediculus humanus capitis, often known as the human head louse, is a blood-sucking ectoparasite that primarily affects kids in both industrialized and developing nations. Permethrin is the primary active component of chemical pediculicides, which are the first line of defense. Despite the prolonged usage of these products, no studies have been conducted to determine if head lice in Honduras are resistant to insecticides. Knockdown resistance (kdr), the most prevalent mechanism in head lice, is caused by two point mutations and the corresponding amino acid substitutions, T917I and L920F, in the voltage-sensitive sodium channel (VSSC) [41]. The most significant contributing factor to the rise in head lice infestations worldwide may be pyrethroids resistance [42, 43]. Knockdown resistance (kdr), which reduces an insect's nerve sensitivity, is a property of lice resistant to pyrethroids and is brought on by single nucleotide point mutations (SNPs) in the para-orthologous voltage-sensitive sodium channel (VSSC) gene. It is well recognized that resistance is caused by the key amino acid substitutions T917I and L920F, which are found in domain II [44]. The locations of the housefly VSSC's amino acid sequence revealed that the mutations T929I and L932F, which have been linked to permethrin resistance, were expressed (rather than in the head louse amino acid sequence). Additionally, it has been shown that this group of mutations coexists as a resistant haplotype; when T197I was produced in *Xenopus oocytes*, either alone or in combination, it effectively inhibited permethrin sensitivity. The T917I amino acid change is relevant to pyrethroid resistance via the kdr-type nerve insensitivity mechanism and can be employed as a molecular marker for resistance detection [45].

11.3 Fruit fly

Cyclodiene and phenylpyrazole insecticides affect the GABA-gated chloride channel component that the resistance to dieldrin gene, or Rdl, encodes. By genetically mapping cyclodiene dieldrin resistance in *Drosophila melanogaster*, the gene was first identified. The change from Ala301 to Ser, one amino acid, caused the 4000-fold resistance. A wide variety of resistant insect species' Rdl orthologs were later found to contain the same alteration. In a research, a duplication at the Rdl gene in *D. melanogaster* was discovered. Rdl is present in two copies, one of which is WT and the other of which has two point mutations: An Ala301 to Ser resistance mutation and a Met360 to Ile substitution. Individuals with this duplication had lower temperature sensitivity, altered RNA editing linked to the resistant allele, and intermediate dieldrin resistance compared to single copy Ser301 homozygotes. This genomic rearrangement is caused by ectopic recombination between Roo transposable elements. By building a transgenic, artificial duplication integrating the 55.7-kb Rdl locus with a Ser301 mutation into an Ala301 background, the duplication phenotypes were confirmed. In most cases, gene duplications increase the amount of gene product generated, which has a considerable impact on the evolution of pesticide resistance. However, in this instance, duplication of the Rdl target site results in permanent heterozygosity, offering a rare opportunity for adaptive mutations to accumulate in a single copy without removing the essential gene's innate function [46].

11.4 Mosquito

The environmental changes in nature and the adaptive genes are easily identifiable; *Culex pipiens* mosquito's resistance to organophosphorus pesticides provides a useful model for analyzing the fitness cost of resistance genes and their origin. This resistance is caused by two loci, the super-locus Ester and the locus Ace.1, each of which contains a number of resistance alleles. According to population surveys, the fitness costs of various resistance genes and even resistance alleles at the same locus vary. The consequences of these resistance genes on various fitness-related variables are being investigated in order to better understand this fitness cost and its unpredictability. The impact of three resistance alleles such as Ester4, Ester1, and Ace.1R on paternity success relative to susceptible males and relative to one another in the research using competition trials between two males for accessing a single female were examined. The impact of susceptible and resistant female genotypes on male mating success was eventually examined. The strains utilized in this investigation have a common genetic history. Males who competed against any of the resistant males had a mating advantage, indicating a high cost of resistance genes for this feature. Regardless of the genotype of the female, resistant male had the same paternity success rate when competing against susceptible males [31, 32, 38, 47].

12. Pathways involved in metabolic resistance

Xenobiotics are detoxified by enzymes into a less or nontoxic compound, resulting in the formation of a more suitable form of metabolite for rapid removal from the body. Insects having resistance metabolize these insecticidal compounds faster due to presence of enzyme with increased catalytic rate and in higher quantities because of increased amplification and transcription of their genes. There are two phases of detoxification: phase I (primary), consisting of oxidation or hydrolysis, and phase II (secondary), consisting of conjugation reactions of products of phase I with different endogenous compounds, like glucuronic acid or glutathione, facilitating their subsequent dissolution and excretion from all over the body [48–51]. Sequestration is also an important mechanism of defense that has been adopted by insects to tolerate these xenobiotics, in addition to such processes of detoxification that are based on cleavage and excretion of insecticides by using enzymes. This strategy involves selective and specific uptake, transportation, and storing of secondary metabolites from the plants on which they are feeding. These metabolites provide them resistance against the insecticides, interfering with their physiological mechanisms [52, 53]. One of the examples of such mechanisms is hematophagy, found in mosquitoes. It could be probably a way of secondary adaptation in which they obtain food of high quality in order to maintain egg production [54].

The enzymes that are majorly involved in xenobiotics detoxification in living organisms are synthesized by transcription of members of large families multigene complexes of enzymes like oxidases, esterases, and glutathione transferases (GSTs).

12.1 Esterases

Esterases belong to a large group of enzymes that catalyze phase 1 reactions, which can metabolize a large variety of endogenous and exogenous substrates. Their role in detoxification of insecticide metabolites is well reported, and they have been shown to act against a wide range of chemical compounds, including organophosphates, pyrethroids, and carbamates [48]. Studies have shown their probable involvement in resistance against Bt toxin [55] and even against neonicotinoid [56]. Insecticide compounds can be detoxified through enzymatic cleavage or sequestration. Insecticides esters are hydrolyzed into their corresponding alcohols and acids by the Esterases and are excreted from the insects' body more easily due to their increased solubility. Insecticides can also be sequestered by Esterases so that the availability of toxic molecules is no longer possible for interacting with the target proteins [57–59]. Esterases are linked to insecticide resistance, due to some qualitative or quantitative or both types of changes in many species of insects, causing the enzymes' overproduction or their structures modifications [48]. Esterases are overexpressed due to upregulation of their genes or amplification or both. One of the most studied examples of detoxification of insecticide through gene amplification is seen in the green peach aphid *Myzus persicae*, which involves the overproduction of a specific enzyme carboxylesterase (Hemiptera: Aphididae) [60–63]. Such amplified esterases have also been seen in mosquitoes of the genus Culex, associated with insecticide resistance (Diptera: Culicidae) [19, 64, 65] and some other species, like the brown planthopper *Nilaparvata lugens* (Stal) (Hemiptera: Delphacidae) [66]. In some species, like *Aphis gossypii* Glover (Hemiptera: Aphididae) or B-biotype *Bemisia tabaci* (Gennadius) (Hemiptera: Aleyrodidae), the expression of enzymes esterases is increased due to increased levels of transcription, due to corresponding gene upregulation [67, 68]. Esterases are also involved in changing the structure of enzymes that are involved in enhanced ability of insects to metabolize the compounds of insecticides. "Mutant ali-esterase theory" was presented by scientists based on this type of mechanism described in the housefly *Musca domestica* (Diptera: Muscidae) for the first time [69]. The insects with resistance exhibited a decreased activity of esterase apparently, as compared to susceptible compounds, resulting due to structural modifications in the enzyme facilitating the process of hydrolysis of the metabolite of insecticide, but it reduced or prevented the hydrolysis of the molecule used for determination of the esterase activity. This mechanism of resistance was considered to be based on substitution of two amino-acid (Gly137Asp and Trp251Leu) in houseflies as well as in many other insect species belonging to the order of Diptera [22, 23, 70].

Other mechanisms like chromosomal rearrangements or demethylation also effect the overproduction of esterase. Mechanisms of demethylation can lead to gene silencing and subsequent reduction of levels of esterase among E4 populations [71], whereas no correlations have been seen between esterase activity and methylation levels in the esterase variant of FE4 [61]. In some Italian populations of aphid species, autosomal rearrangements of same type for the FE4 isoforms of esterase have been reported but only in those with low activity of esterase, showing that resistance due to esterase and translocation are not correlated always [62, 72].

12.2 Monooxygenases

Another class of enzymes involved in metabolism of Xenobiotics is microsomal oxidases or mixed function oxidases (MFOs). These are enzymes of phase 1 reactions and are also involved in metabolism of endogenous metabolites like fatty acids, pheromones, or hormones. These enzymes can convert hydrophobic molecules to hydrophilic substances so that they can easily be eradicated from the body. Their major localization is in digestive tract [37, 73, 74]. Microsomal oxidases are Cytochrome P450 monooxygenases (P450s) that are from the group of enzymes composed of heme thiolate proteins. They exhibit a characteristic peak of absorbance at 450 nm when they are in a reduced form and complexed with molecule of carbon monoxide. The reactions catalyzed by these enzymes involve the transfer of one atom of molecular oxygen to a substrate and the reduction of the second atom of oxygen to form water. This process needs the transfer of at least two electrons, which are provided by NADPH cytochrome P450 reductase [73, 75]. P450s possess a large variety of enzymes that are highly specific for substrate and can catalyze various reactions like hydroxylation, epoxidation, desulfurization, O-dealkylation, or N-dealkylation. They play a major role in interactions between plants and insects and metabolizing many insecticides like organophosphates, carbamates, neonicotinoids, and pyrethoids [50, 76–79]. The enzymes of P450 family are named by abbreviation CYP with an Arabic number of the respective family, a capital letter designating the subfamily with an Arabic numeral designating the individual protein. Every each has its own gene to be coded. More than 600 P450 genes have been characterized from insects, and it has been found that genes of families CYP6, CYP4, CYP12, and CYP9 are associated with resistance in insects against insecticides (**Figures 1** and **2**) [48, 73, 74]. The MPOs are majorly found in the midgut, Malpighian tubules, and fat bodies of insects. The housefly is the candidate whose MPOs system has been extensively studied [80]. Studies have reported that higher concentrations of P450s and an increased activity of monooxygenases are found in resistant insects. Overexpression of such activities occurs as a result of upregulation of their genes, that is mediated by the modifications of regulatory elements [73]. They have also shown amplification of genes or qualitative modifications in other studies [78, 81–83]. Some type of insecticides can also be activated by enzymes of insect P450 system. One of the example is the formation of phosphate (P=O) from phosphorothioates (P=S). This causes an increased 330 potency for inhibiting acetylcholinesterase by a magnitude of 3 or 4 orders. The synthesis of juvenile hormone, pheromone components, and ecdysone also needs involvement of P450s [84].

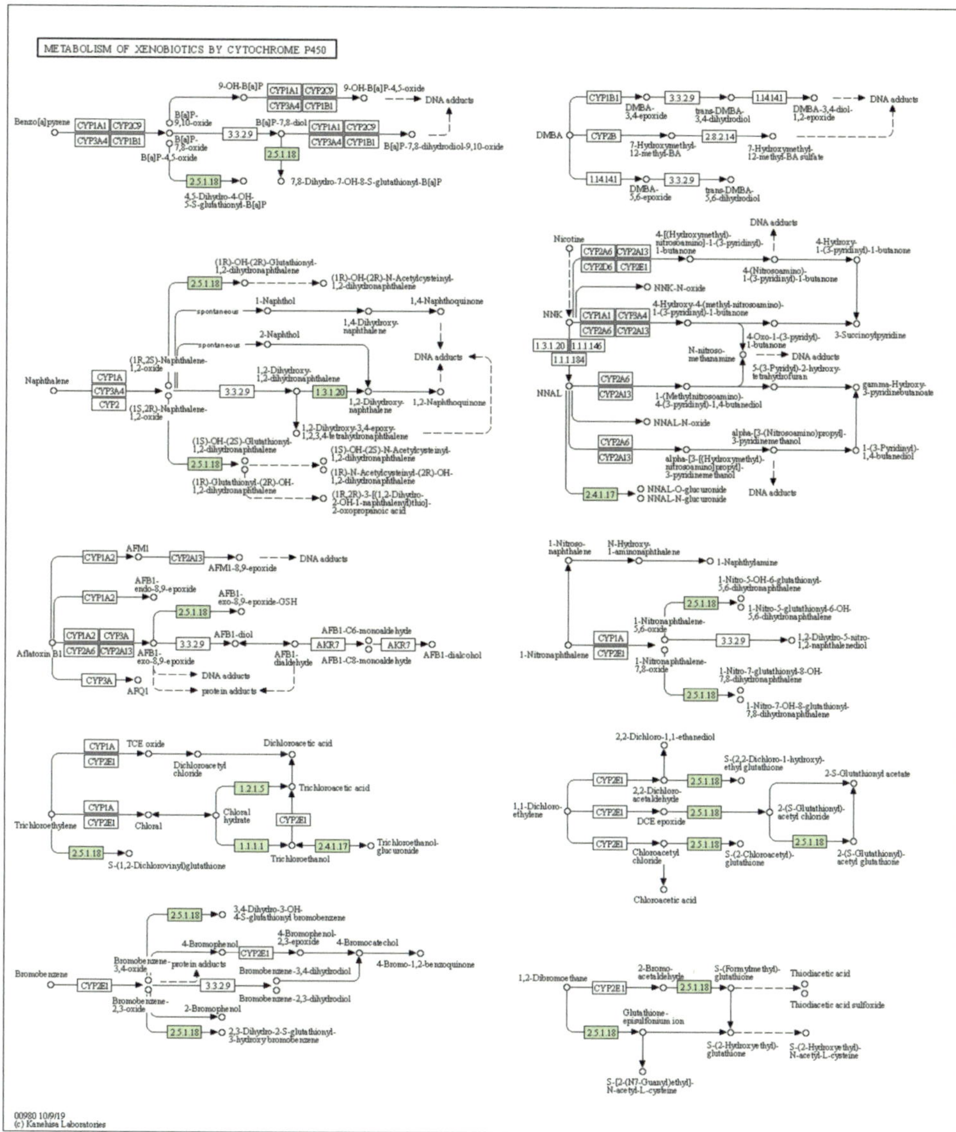

Figure 1.
Metabolism of xenobiotics by cytochrome P450 - Anopheles gambiae (malaria mosquito).

Aerobic organisms are characterized by the presence of a diverse group or family of enzymatic proteins named Glutathione transferases (GSTs), which are found ubiquitously. They play a major function in the detoxification of xenobiotic as well as endogenous compounds. They are also found to be associated with synthesis of hormones, intracellular transport, and protection in contradiction of oxidative stress [85]. GST enzymes are involved in catalyzing the conjugation reactions of reduced glutathione with electrophilic molecules or substrates as well as in sequestrating substrate. This results in increased hydrophilicity or water solubility and decreased toxicity of reactive molecules and in turn facilitates their removal or excretion from the body. Specifically, they catalyze conjugations by facilitating nucleophilic attack of the sulfhydryl group of endogenous reduced glutathione (GSH) on electrophilic centers of a range of xenobiotic

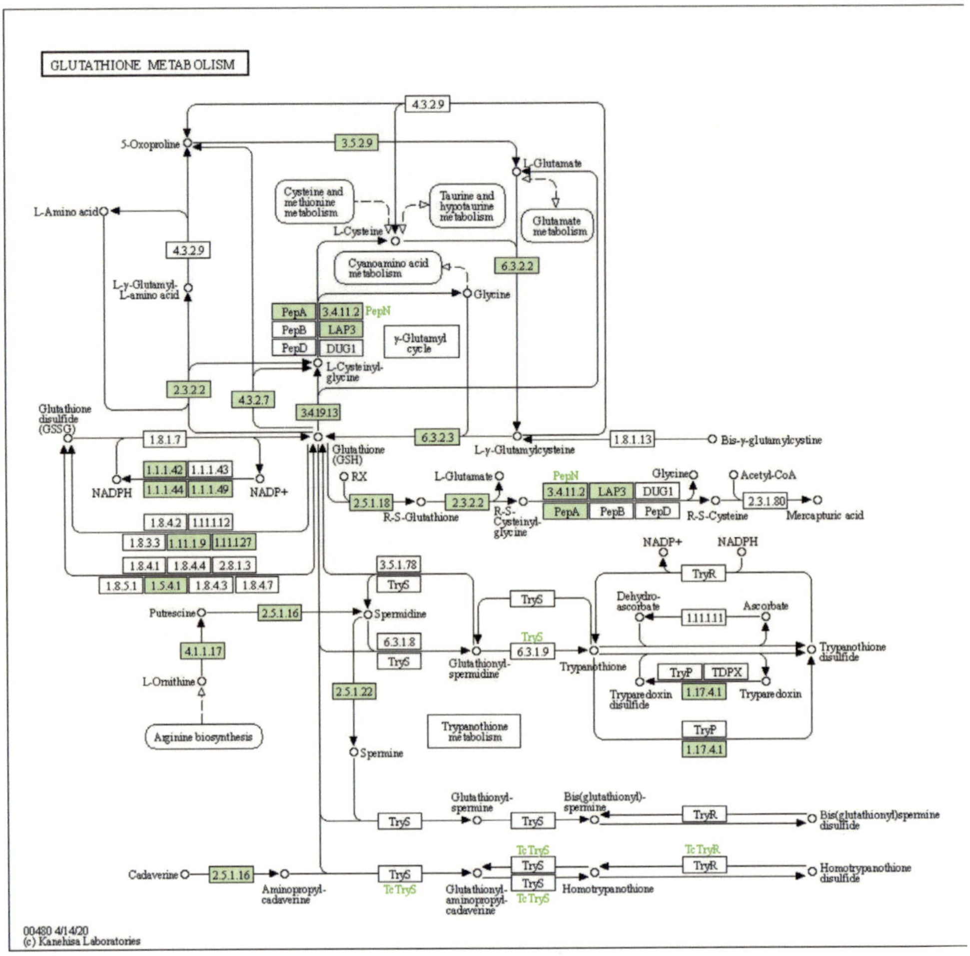

Figure 2.
Glutathione metabolism - Drosophila melanogaster (fruit fly).

compounds, including insecticides or acaricides [86] and various plant toxins [76]. This results in the conversion of xenobiotics into derivatives of mercapturic acid, which are more soluble and easily excreted out from the insect body [87, 88]. Species of free oxygen radicals are formed in insects by the action of pesticides that are highly toxic and can be removed with the help of these GSTs. They also help in metabolizing insecticides through their reductive dehydrochlorination [89]. In insects, there are two groups of GSTs: cytosolic and microsomal. They are classified on the basis of their occurrence in the cell. Only cytosolic GSTs have been reported to be involved in insecticides metabolism. They play an important role in developing resistance against some insecticides like pyrethroids and organophosphates. In mosquitoes and houseflies, resistance against DDT has been eveloved due to a DDT dehydrochlorinase GST [88]. The full extent of family of this enzyme has been revealed in genomes of the *Drosophila melanogaster* Meigen and *Anopheles gambiae* Giles [88]. Generally, the quantity of enzyme is increased due to either overexpression of gene or its amplification, which results in enhanced resistance on the basis of GSTs [90, 91]. Insecticides can also be sequestered by GSTs that provide the insects protection against the toxicity of these insectcides, for example, pyrethroid [83].

12.3 Pgp pumps

Pgp pumps are transporters composed of P-glycoprotein (Pgp) that are integral membrane proteins and belong to the ATP-binding cassette (ABC) superfamily, which utilizes the energy produced from ATP breakdown and translocates different metabolites as well as xenobiotics across the cell membranes [92]. This type of mechanism has been majorly observed in fungi and bacteria for developing resistance against antibiotics [93], but very little work has been reported on it regarding insects. Only recently, these ABC transporters have been found in insects as a supposed mechanism that can contribute in resistance by facilitating the efflux transport mechanism of insecticides as well as their compounds or metabolites that are derived from phase I and II reactions [94–98]. ABC transporters can produce resistance in insects through different modes like quantification of protein or transcript and by synergistic mechanisms of ABC inhibitors [94, 99]. Furthermore, in different lepidopteran species, a mutant allele has been discovered that confers resistance to the pore-forming Cry1Ac toxin from *Bacillus thuringiensis* (Bt) by a mechanism that is not related to toxin extrusion, but because it causes the loss of Cry1Ac binding to membrane vesicles [100, 101].

13. What is systems biology?

The study of the relationships and behavior of biological entity components such as molecules, cells, organs, and organisms is known as systems biology. Individual roles are played by microbes, plants, animals, and entire ecosystems in the natural world, which is a complex system of interconnected pieces. The investigation of living creatures is approached comprehensively in systems biology. It studies how diverse biological creatures interact at different sizes. Every person, for example, is a system. The system includes our organs, tissues, cells, and the components they are formed of, as well as bacteria and other creatures that dwell on our epidermis and in our digestive system [102].

Computational and mathematical analysis and modeling are important to systems biology. It gathers data from a wide variety of biological sciences and technologies known as "-omics" by researchers. Among these "omics" are genomics (the study of whole gene sets in an organism) and proteomics (the study of all the proteins in a cell, tissue, or organism). The emphasis in these fields is on describing and measuring the biological molecules that underpin how organisms are produced, operate, and live [103].

14. What is a significant role of systems biology in causing insect resistant?

Insecticide resistance is regarded as a typical similar pattern of microevolution, in which a powerful selection agent is given to a large natural community, resulting in a shift in the frequency of alleles conferring resistance. While numerous pesticide resistance variations have been identified at the gene level that was in term of systems biology, they are usually single genes with a big influence seen in highly resistant insect pest. With *Drosophila melanogaster,* many polymorphisms have been involved in DDT resistance; however, only Cyp6g1 locus has already been proven to be meaningful to field populations. They uncover DDT-associated polygenes using genome-wide association studies (GWAS) and assess their adaptive importance using selective sweep analysis. As a result, they validate two DDT resistance loci. This was considered as the significant role of system biology in causing insect resistant [104].

15. What are the main pathways involved in insects resistance?

The two major pathways involved in insecticide resistance were metabolic resistance and target-site resistance. Metabolic resistance is a typical defense strategy that relies on enzymatic mechanisms to protect the insect by detoxifying/sequestering pesticide compounds. In order to overcome the potential toxicity of the plants they feed on, the enzymes involved are those that insects have evolved as support against naturally occurring plant poisons (study will focus) such as alkaloids, terpenes, and phenols. This might explain the modernization of metabolic resistance to a wide range of insecticides, many of which have direct or indirect botanical origins. Enzymes may detoxify xenobiotics into a nontoxic chemical and/or a form that is more suited for fast removal from the body [105].

Resistant insects metabolize the pesticide quicker because they have enzymes with a better catalytic rate, or because they have more enzymes as a result of enhanced transcription or gene duplication. Detoxification can be separated into two phases: phase I (primary) activities involving hydrolysis or oxidation, and phase II (secondary) processes involving coupling of phase I results with endogenous molecules such as glutathione and eventual elimination from the body. In addition to such enzymatic cleavage and excretion-based detoxification methods, sequestration is a significant defense mechanism that certain insects have evolved to withstand xenobiotics [67].

This is a typical phenomenon in insect herbivores that involves the precise and selective absorption, transport, and storing of secondary metabolites from plants in order to avoid interference with the insects' physiological processes. Such behavior has been seen in mosquitos, where hematophagy is most likely a subsequent adaption to get high-quality food for egg formation. Members of vast multigene families of isoenzymes, oxidoreductases, and GSTs transcribe the enzymes involved in xenobiotic detoxification in living organisms [106].

On the other hand, target-site resistance was explained as the pesticide's target site of action in the insect that can be genetically engineered to inhibit the insecticide from bonding or interacting at the site of action, lowering or eliminating the insecticide's pesticidal impact. During target-site resistance, an insecticide's particular binding site is transformed (mutated) and/or removed, rendering the target site unsuitable with activation. Most frequent insect (*Myzus persicae*, *Musca domestica*, and *Drosophila melanogaster*) target areas are mutated, including subunits such as cholinergic acetyl cholinergic receptor (nAChRs), knockdown resistance (KDR), and others. Insecticides are not able to bind inside the target area as a result of these changes, resulting in a reduction of binding affinity [107].

16. Which mechanism of resistance affects the behavior of the insects?

Metabolic resistance serves as the most common mode and frequently poses the most difficult barrier. Insects break down pesticides using their internal enzyme systems. These enzymes may be present in larger concentrations or in more effective forms in resistant strains. It was also explained by the case study of P450 gene in *House Flies* [108].

Insects may employ a variety of metabolic processes to avoid the fatal effects of pesticides. Increased cytochrome P450 detoxification, for example, is known to play a key role in many insect species. P450s' constitutively elevated overexpression and induction are hypothesized to somehow be responsible for enhanced levels of pesticide detoxification. However, unlike continuously upregulation P450 genes, whose regulation connection with pesticide resistance has been well explored; P450 induction in insecticide resistance is less well understood. The current work focuses on the identification of particular P450 genes that are activated in permethrin-resistant house flies in response to permethrin treatment. As a result, Permethrin administration co-upregulated the expression of three P450 genes, CYP4D4v2, CYP4G2, and CYP6A38, in permethrin conferring resistance ALHF house flies in a period and dose-dependent way. The protein sequences among these 3 P450s from resistant ALHF as well as vulnerable aabys and CS house flies were found to be similar. CYP4D4v2 and CYP6A38 were found on autosome 5, correlating to the association of P450-mediated resistance in ALHF, while CYP4G2 was found on autosome 3, where the key insecticide susceptibility factors for ALHF had been mapped, but no P450 genes had been reported previously to this investigation.

This study provided the first direct proof that numerous P450 genes are co-upregulated in permethrin-resistant house flies via the induction process, which boosts total P450 gene expression levels in resistant house flies. This research provides new information on the functional importance of P450 genes as they react to insecticide therapies, detoxification of insecticides, insect adaptation to their atmosphere, and the evolution of insects [108].

17. What is the role of protein-protein interaction pathway in insect resistance according to system biology?

At the moment, the problem of resistance is not fundamentally solved since the development speed of new insecticides cannot keep up with the progression speed of resistance, and there is a lack of knowledge of the molecular mechanism of resistance.

Researchers used literature mining and the String database to identify seed genes and their interacting proteins involved in the biological mechanism of pesticide resistance in *Drosophila melanogaster*. They discovered 528 proteins molecules and 13,514 protein-protein interactions. String and Pajek built the protein interaction network, and we looked at topological features like degree centrality and eigenvector centrality. KEGG pathway enrichment analyses revealed an enrichment for proteasome complexes and drug metabolism of cytochrome P450. This is the first time that the pesticide resistance in molecular level mechanism of *D. melanogaster* has been investigated using network biology methodologies and tools, and it can provide a bioinformatic basis for further understanding of insecticide resistance mechanisms [108].

So, systems biology plays a very important role in the insect resistance against insecticides and different chemicals such as DDT and permethrin. From systems biology, not only the identification of genes was done, but also the protein-protein interactions were found out, which were responsible for insect resistance.

Author details

Rabbiah Manzoor Malik[1,2,3*], Sahar Fazal[3], Narjis Khatoon[3], Muneeba Ishtiaq[3], Saima Batool[3] and Syed Tauqeer Abbas[1,2]

1 Wah Medical College, Wah Cantt, Pakistan

2 National University of Medical Sciences, Rawalpindi, Pakistan

3 Capital University of Sciences and Technology, Islamabad, Pakistan

*Address all correspondence to: rabbiahmanzoor@gmail.com

References

[1] Van Huis A. Potential of insects as food and feed in assuring food security. Annual Review of Entomology. 2013;**58**:563-583

[2] Pietro Campobasso C, Di Vella G, Introna F. Factors affecting decomposition and Diptera colonization. Forensic Science International. 2001;**120**(1-2):18-27

[3] Mueller UG, Gerardo NM, Aanen DK, Six DL, Schultz TR. The evolution of agriculture in insects. Annual Review of Ecology, Evolution, and Systematics. 2005;**36**(1):563-595

[4] Wang Y-S, Shelomi M. Review of black soldier fly (Hermetia illucens) as animal feed and human food. Food. 2017;**6**(10):91

[5] Nghiem LTP et al. Economic and environmental impacts of harmful non-indigenous species in Southeast Asia. PLoS One. 2013;**8**(8):e71255

[6] Gregory PJ, Johnson SN, Newton AC, Ingram JSI. Integrating pests and pathogens into the climate change/food security debate. Journal of Experimental Botany. 2009;**60**(10):2827-2838

[7] Carvalho FP. Agriculture, pesticides, food security and food safety. Environmental Science and Policy. 2006;**9**(7-8):685-692

[8] Delaunay P et al. Bedbugs and infectious diseases. Clinical Infectious Diseases. 2011;**52**(2):200-210

[9] Ware GW, Whitacre DM. An introduction to insecticides. Pesticide Biochemistry. 2004;**6**

[10] Perry T, Batterham P, Daborn PJ. The biology of insecticidal activity and resistance. Insect Biochemistry and Molecular Biology. 2011;**41**(7):411-422

[11] Liu N. Insecticide resistance in mosquitoes: Impact, mechanisms, and research directions. Annual Review of Entomology. 2015;**60**:537-559

[12] Ranson H, Burhani J, Lumjuan N, Black WC IV. Insecticide resistance in dengue vectors. Trop. net [online]. 2010;**1**(1)

[13] Ellegren H. Genome sequencing and population genomics in non-model organisms. Trends in Ecology & Evolution. 2014;**29**(1):51-63

[14] Allendorf FW, Hohenlohe PA, Luikart G. Genomics and the future of conservation genetics. Nature Reviews. Genetics. 2010;**11**(10):697-709

[15] Brown TA. Genomes 4. Garland Science. 2018. pp. 1-24

[16] Ffrench-Constant RH. The molecular genetics of insecticide resistance. Genetics. 2013;**194**(4):807-815

[17] Ranson H et al. Evolution of supergene families associated with insecticide resistance. Science (80-.). 2002;**298**(5591):179-181

[18] Daborn PJ et al. A single P450 allele associated with insecticide resistance in drosophila. Science (80-.). 2002;**297**(5590):2253-2256

[19] Raymond M, Chevillon C, Guillemaud T, Lenormand T, Pasteur N. An overview of the evolution of overproduced esterases in the mosquito Culex pipiens. Philosophical Transactions of the Royal Society London Series B Biological Sciences. 1998;**353**(1376):1707-1711

[20] Field LM, Devonshire AL, Forde BG. Molecular evidence that insecticide resistance in peach-potato aphids (Myzus persicae Sulz.) results from amplification of an esterase gene. The Biochemical Journal. 1988;**251**(1):309-312

[21] Newcomb RD, Campbell PM, Ollis DL, Cheah E, Russell RJ, Oakeshott JG. A single amino acid substitution converts a carboxylesterase to an organophosphorus hydrolase and confers insecticide resistance on a blowfly. Proceedings of the National Academy of Sciences. 1997;**94**(14):7464-7468

[22] Claudianos C, Russell RJ, Oakeshott JG. The same amino acid substitution in orthologous esterases confers organophosphate resistance on the house fly and a blowfly. Insect Biochemistry and Molecular Biology. 1999;**29**(8):675-686

[23] Campbell PM, Newcomb RD, Russell RJ, Oakeshott JG. Two different amino acid substitutions in the ali-esterase, E3, confer alternative types of organophosphorus insecticide resistance in the sheep blowfly, Lucilia cuprina. Insect Biochemistry and Molecular Biology. 1998;**28**(3):139-150

[24] French-Constant RH, Anthony N, Aronstein K, Rocheleau T, Stilwell G. Cyclodiene insecticide resistance: From molecular to population genetics. Annual Review of Entomology. 2000;**45**(1):449-466

[25] Williamson MS, Martinez-Torres D, Hick CA, Devonshire AL. Identification of mutations in the houseflypara-type sodium channel gene associated with knockdown resistance (kdr) to pyrethroid insecticides. Molecular and General Genetics MGG. 1996;**252**(1):51-60

[26] Ibrahim MA, Griko N, Junker M, Bulla LA. Bacillus thuringiensis: A genomics and proteomics perspective. Bioengineering Bugs. 2010;**1**(1):31-50

[27] Olazcuaga L et al. A whole-genome scan for association with invasion success in the fruit fly Drosophila suzukii using contrasts of allele frequencies corrected for population structure. Molecular Biology and Evolution. 2020;**37**(8):2369-2385

[28] Homem RA, Davies TGE. An overview of functional genomic tools in deciphering insecticide resistance. Current Opinion in Insect Science. 2018;**27**:103-110

[29] Jin T, Zeng L, Lin Y, Lu Y, Liang G. Insecticide resistance of the oriental fruit fly, Bactrocera dorsalis (Hendel) (Diptera: Tephritidae), in mainland China. Pest Management Science. 2011;**67**(3):370-376

[30] Ocampo CB, Salazar-Terreros MJ, Mina NJ, McAllister J, Brogdon W. Insecticide resistance status of Aedes aegypti in 10 localities in Colombia. Acta Tropica. 2011;**118**(1):37-44

[31] Faucon F et al. Identifying genomic changes associated with insecticide resistance in the dengue mosquito Aedes aegypti by deep targeted sequencing. Genome Research. 2015;**25**(9):1347-1359

[32] Yin J et al. Molecular detection of insecticide resistance mutations in Anopheles gambiae from Sierra Leone using multiplex SNaPshot and sequencing. Frontiers in Cellular and Infection Microbiology. 2021;**11**:778

[33] Karaa SU. Insecticide resistance. London, UK: IntechOpen; 2012

[34] Balabanidou V, Grigoraki L, Vontas J. Insect cuticle: A critical determinant of insecticide resistance. Current Opinion in Insect Science. 2018;**27**:68-74

[35] Zalucki MP, Furlong MJ. Behavior as a mechanism of insecticide resistance: Evaluation of the evidence. Current Opinion in Insect Science. 2017;**21**:19-25

[36] Ilias A, Vontas J, Tsagkarakou A. Global distribution and origin of target site insecticide resistance mutations in Tetranychus urticae. Insect Biochemistry and Molecular Biology. 2014;**48**:17-28

[37] Liu N, Li M, Gong Y, Liu F, Li T. Cytochrome P450s--their expression, regulation, and role in insecticide resistance. Pesticide Biochemistry and Physiology. 2015;**120**:77-81

[38] Stevenson BJ, Pignatelli P, Nikou D, Paine MJI. Pinpointing P450s associated with pyrethroid metabolism in the dengue vector, Aedes aegypti: Developing new tools to combat insecticide resistance. PLoS Neglected Tropical Diseases. 2012;**6**(3):e1595

[39] Yanola J, Somboon P, Walton C, Nachaiwieng W, Prapanthadara L. A novel F1552/C1552 point mutation in the Aedes aegypti voltage-gated sodium channel gene associated with permethrin resistance. Pesticide Biochemistry and Physiology. 2010;**96**(3):127-131

[40] Dong KE, Scott JG. Linkage of kdr-type resistance and the Para-homologous sodium channel gene in German cockroaches (Blattella germanica). Insect Biochemistry and Molecular Biology. 1994;**24**(7):647-654

[41] Larkin K et al. First evidence of the mutations associated with pyrethroid resistance in head lice (Phthiraptera: Pediculidae) from Honduras. Parasites & Vectors. 2020;**13**(1):1-7

[42] Falagas ME, Matthaiou DK, Rafailidis PI, Panos G, Pappas G. Worldwide prevalence of head lice. Emerging Infectious Diseases. 2008;**14**(9):1493-1494

[43] Ichard R, Oberts JR. Clinical Practice Head Lice. 2002 [Online]. Available: www.nejm.org [Accessed 16 February 2019]

[44] Karakuş M, Atıcı T, Karabela ŞN, Baylan O, Limoncu ME, Balcıoğlu İC. Detection of permethrin resistance and phylogenetic clustering of turkish head lice (*Pediculus humanus capitis*; De Geer, 1767 populations). Acta Tropica. 2020;**204**:105362

[45] Eldefrawi ME, Eldefrawi AT. Nervous-system-based insecticides. In: Safer Insecticides. London, UK: CRC Press; 2020. pp. 155-207

[46] Dong K et al. Molecular biology of insect sodium channels and pyrethroid resistance. Insect Biochemistry and Molecular Biology. 2014;**50**:1-17

[47] Paul A, Harrington LC, Zhang L, Scott JG. Insecticide resistance in Culex pipiens from New York. Journal of the American Mosquito Control Association. 2005;**21**(3):305-309

[48] Li X, Schuler MA, Berenbaum MR. Molecular mechanisms of metabolic resistance to synthetic and natural xenobiotics. Annual Review of Entomology. 2007;**52**:231-253

[49] Hollingworth RM, Dong K. The biochemical and molecular genetic basis of resistance to pesticides in arthropods. In: Whalon ME, MotaSanchez D, Hollingworth RM, editors. Global Pesticide Resistance in Arthropods. Wallingford, UK: CABI; 2008. pp. 5-31

[50] Yu SJ. The Toxicology and Biochemistry of Insecticide. Boca Raton, FL, USA: CRC Press; 2008. p. XVI+276

[51] Berenbaum MR, Johnson RM. Xenobiotic detoxification pathways in honey bees. Current Opinion in Insect Science. 2015;**10**:51-58

[52] Erb M, Robert CAM. Sequestration of plant secondary metabolites by insect herbivores: Molecular mechanisms and ecological consequences. Current Opinion in Insect Science. 2016;**14**:8-11

[53] Petschenka G, Agrawal AA. How herbivores coopt plant defenses: Natural selection, specialization, and sequestration. Current Opinion in Insect Science. 2016;**14**:17-24

[54] Moore SJ. Plant-based insect repellents. In: Debboun M, Frances SP, Strickman DA, editors. Insect Repellents Handbook. Boca Raton, FL, USA: CRC Press; 2015. pp. 179-212

[55] Gunning RV, Dang HT, Kemp FC, Nicholson IC, Moores GD. New resistance mechanism in Helicoverpa armigera threatens transgenic crops expressing bacillus thuringiensis Cry1Ac toxin. Applied and Environmental Microbiology. 2005;**71**:2558-2563

[56] Zhu YC, Luttrell R. Altered gene regulation and potential association with metabolic resistance development to imidacloprid in the tarnished plant bug, Lygus lineolaris. Pest Management Science. 2015;**71**:40-57

[57] Devonshire AL, Moores GD. A carboxylesterase with broad substrate specificity causes organophosphorus, carbamate and pyrethroid resistance in peach-potato aphids (Myzus 333 persicae). Pesticide Biochemistry and Physiology. 1982;**18**:235-246

[58] Oakeshott JG, Claudianos C, Campbell PM, Newcomb RD, Russell RJ. Biochemical genetics and genomics of insect esterases. In: Gilbert LI, Iatrou K, Gill SS, editors. Comprehensive Insect Molecular Science. Vol. 5. Oxford, UK: Elsevier; 2005. pp. 309-381

[59] Wheelock CE, Shan G, Ottea J. Overview of carboxylesterases and their role in the metabolism of insecticide. Journal of Pest Science. 2005;**30**:75-83

[60] Field LM, Devonshire AL, Forde BG. Molecular evidence that insecticide resistance in peachpotato aphids (Myzus persicae Sulz.) results from amplification of an esterase gene. The Biochemical Journal. 1988;**251**:309-312

[61] Bizzaro D, Mazzoni E, Barbolini E, Giannini S, Cassanelli S, Pavesi F, et al. Relationship among expression, amplification and methylation of FE4 esterase genes in Italian populations of Myzus persicae (Sulzer) (Homoptera: Aphididae). Pesticide Biochemistry and Physiology. 2005;**81**:51-58

[62] Rivi M, Monti V, Mazzoni E, Cassanelli S, Panini M, Anaclerio M, et al. A1-3 chromosomal translocations in Italian populations of the peach potato aphid Myzus persicae (Sulzer) not linked to esterase-based insecticide resistance. Bulletin of Entomological Research. 2013;**103**:278-285

[63] Bass C, Puinean M, Zimmer TC, Denholm I, Field LM, Foster SP, et al. The evolution of insecticide resistance in the peach potato aphid, *Myzus persicae*. Insect Biochemistry and Molecular Biology. 2014;**51**:41-51

[64] Severini C, Marinucci M, Raymond M. Insecticide resistance genes in Culex pipiens (Diptera: Culicidae) from Italy: Esterase B locus at the DNA level. Journal of Medical Entomology. 1994;**3**:496-499

[65] Hemingway J, Hawkes NJ, McCarroll L, Ranson H. The molecular basis of insecticide resistance in mosquitoes. Insect Biochemistry and Molecular Biology. 2004;**34**:653-665

[66] Small GJ, Hemingway J. Molecular characterization of the amplified

carboxylesterase gene associated with organophosphorus insecticide resistance the brown planthopper, *Nilaparvata lugens*. Insect Molecular Biology. 2000;**9**:647-653

[67] Alon M, Alon F, Nauen R, Morin S. Organophosphates' resistance in the B-biotype of Bemisia tabaci (Hemiptera: Aleyrodidae) is associated with a point mutation in an ace1- type acetylcholinesterase and overexpression of carboxylesterases. Insect Biochemistry and Molecular Biology. 2008;**38**:940-949

[68] Cao CW, Zhang J, Cao XW, Liang P, Cuo HL. Overexpression of carboxylesterase gene associated with organophosphorous insecticide resistance in cotton aphids, Aphis gossypii (glover). Pesticide Biochemistry and Physiology. 2008;**90**:175-180

[69] Oppenoorth FJ, van Asperen K. Allelic genes in the housefly producing modified enzymes that cause organophosphate resistance. Science. 1960;**132**:298-299

[70] Carvalho RA, Torres T, Azeredo-Espin AML. A survey of mutations in the Cochliomyia hominivorax (Diptera: Calliphoridae) esterase E3 gene associated with organophosphate resistance and the molecular identification of mutant alleles. Veterinary Parasitology. 2006;**140**:344-351

[71] Field LM. Methylation and expression of amplified esterase genes in the aphid Myzus persicae (Sulzer). The Biochemical Journal. 2000;**349**:863-868

[72] Rivi M, Monti V, Mazzoni E, Cassanelli S, Panini M, Bizzaro D, et al. Karyotype variations in Italian populations of the peach-potato aphid Myzus persicae (Hemiptera: Aphididae). Bulletin of Entomological Research. 2012;**102**:663-671

[73] Feyereisen R. Insect cytochrome P450. In: Gilbert LI, Iatrou K, Gill SS, editors. Comprehensive Molecular Insect Science. Vol. 4. Oxford: Elsevier BV; 2005. pp. 1-77

[74] Feyereisen R. Insect P450 inhibitors and insecticides: Challenges and opportunities. Pest Management Science. 2015;**71**:793-800

[75] Guengerich FP. Common and uncommon cytochrome P450 reactions related to metabolism and chemical toxicity. Chemical Research in Toxicology. 2008;**14**:611

[76] Després L, David JP, Gallet C. The evolutionary ecology of insect resistance to plant chemicals. Trends in Ecology & Evolution. 2007;**22**:298-307

[77] Philippou D, Field L, Moores G. Metabolic enzyme(s) confer imidacloprid resistance in a clone of Myzus persicae (Sulzer) (Hemiptera: Aphididae) from Greece. Pest Management Science. 2010;**66**:390-395

[78] Puinean AM, Foster SP, Oliphant L, Denholm I, Field LM, Millar NS, et al. Amplification of a cytochrome P450 gene is associated with resistance to neonicotinoid insecticides in the aphid Myzus persicae. PLoS Genetics. 2010;**6**:e1000999

[79] Alptekin S, Bass C, Nicholls C, Paine MJ, Clark SJ, Field L, et al. Induced thiacloprid insensitivity in honeybees (Apis mellifera L.) is associated with up-regulation of detoxification genes. Insect Molecular Biology. 2016;**25**:171-180

[80] Markussen MDK, Kristensen M. Cytochrome P450 monooxygenase-mediated neonicotinoid resistance in the house fly Musca domestica L. Pesticide Biochemistry and Physiology. 2010;**98**:50-58

[81] Amichot M, Tarés S, Brun-Barale A, Arthaud L, Bride JM, Berge JB. Point mutations associated with insecticide resistance in the drosophila cytochrome P450 CYP6A2 enable DDT metabolism. European Journal of Biochemistry. 2004;**271**:1250-1257

[82] Wondji SC, Irving H, Morgan J, Lobo NF, Collins FH, Hunt RH, et al. Two duplicated P450 genes are associated with pyrethroid resistance in Anopheles funestus, a major malaria vector. Genome Research. 2009;**19**:452-459

[83] Kostaropoulos I, Papadopoulos AI, Metaxakis A, Boukouvala E, Papadopoulou-Mourkidou E. Glutathione S-transferase in the defence against pyrethroids in insects. Insect Biochemistry and Molecular Biology. 2001;**31**:313-319

[84] Feyereisen R. Molecular biology of insecticide resistance. Toxicology Letters. 1995;**82**:83-90

[85] Ketterman AJ, Saisawang C, Wongsantichon J. Insect glutathione transferases. Drug Metabolism Reviews. 2011;**43**:253-265

[86] Konanz S, Nauen R. Purification and partial characterization of a glutathione S-transferase from the two-spotted spider mite, Tetranychus 334 urticae. Pesticide Biochemistry and Physiology. 2004;**79**:49-57

[87] Habig WH, Pabst MJ, Jakoby WB. Glutathione Stransferases. The first enzymatic step in mercapturic acid formation. The Journal of Biological Chemistry. 1974;**249**:7130-7139

[88] Enayati AA, Ranson H, Hemingway J. Insect glutathione transferases and insecticide resistance. Insect Molecular Biology. 2005;**14**:3-8

[89] Hayes JD, Flanagan JU, Jowsey IR. Glutathione transferases. Annual Review of Pharmacology and Toxicology. 2005;**45**:51-88

[90] Ranson H, Hemingway J. Glutathione transferases. In: Gilbert LI, Iatrou K, Gill SS, editors. 335 Comprehensive Molecular Insect Science. Oxford: Elsevier; 2005. pp. 383-402

[91] Vontas JG, Small GJ, Nikou DC, Ranson H, Hemingway J. Purification, molecular cloning and heterologous expression of a glutathione S-transferase involved in insecticide resistance from the rice brown planthopper, Nilaparvata lugens. The Biochemical Journal. 2002;**362**:329-337

[92] Hollenstein K, Dawson RJP, Locher KP. Structure and mechanism of ABC transporter proteins. Current Opinion in Structural Biology. 2007;**17**:412-418

[93] Lage H. ABC transporters: Implications on drug resistance from microorganisms to human cancers. International Journal of Antimicrobial Agents. 2003;**22**:188-199

[94] Dermauw W, Van Leeuwen T. The ABC gene family in arthropods: Comparative genomics and role in insecticide transport and resistance. Insect Biochemistry and Molecular Biology. 2014;**45**:89-110

[95] Bariami V, Jones CM, Poupardin R, Vontas J, Ranson H. Gene amplification, ABC transporters and cytochrome P450s: Unravelling the molecular basis of pyrethroid resistance in the dengue vector, *Aedes aegypti*. PLOS Neglected Tropical Diseases. 2012;**6**:e1692

[96] Aurade RM, Jayalakshmi SK, Sreeramulu K. Pglycoprotein ATPase from the resistant pest, Helicoverpa

armigera: Purification, characterization and effect of various insecticides on its transport function. Biochimica et Biophysica Acta. 2010;**1798**:1135-1143

[97] Porretta D, Gargani M, Bellini R, Medici A, Punelli F. Defence mechanisms against insecticides temephos and diflurobenzuron in the mosquitos Aedes caspius: The P-glycoprotein efflux pumps. Medical and Veterinary Entomology. 2008;**22**:48-54

[98] O'Donnell MJ. Insect excretory mechanisms. Advances in Insect Physiology. 2008;**35**:1-122

[99] Buss DS, Callaghan A. Interaction of pesticides with p-glycoprotein and other ABC proteins: A survey of the possible importance to insecticide, herbicide and fungicide resistance. Pesticide Biochemistry and Physiology. 2008;**90**:141-153

[100] Heckel DG. Learning the ABCs of Bt: ABC transporters and insect resistance to bacillus thuringiensis provide clues to a crucial step in toxin mode of action. Pesticide Biochemistry and Physiology. 2012;**104**:103-110

[101] Gahan LJ, Pauchet Y, Vogel H, Heckel DG. An ABC transporter mutation is correlated with insect resistance to bacillus thuringiensis Cry1Ac toxin. PLoS Genetics. 2010;**6**:e1001248

[102] Babtie AC, Kirk P, Stumpf MP. Topological sensitivity analysis for systems biology. In: Proceedings of the National Academy of Sciences. 2014;**111**(52):18507-18512

[103] Karr JR, Sanghvi JC, Macklin DN, Gutschow MV, Jacobs JM, Bolival B Jr, et al. A whole-cell computational model predicts phenotype from genotype by system biology. Cell. 2012;**150**(2):389-401

[104] Duneau D, Sun H, Revah J, San Miguel K, Kunerth HD, Caldas IV, et al. Signatures of insecticide selection in the genome of *Drosophila melanogaster*. G3: Genes, Genomes, Genetics. 2018;**8**(11):3469-3480

[105] Panini M, Manicardi GC, Moores GD, Mazzoni E. An overview of the main pathways of metabolic resistance in insects. Invertebrate Survival Journal. 2016;**13**(1):326-335

[106] Dang K, Doggett SL, Veera Singham G, Lee CY. Insecticide resistance and resistance mechanisms in *Myzus persicae, Musca domestica, and Drosophila melanogaster*. Parasites & Vectors. 2017;**10**(1):1-31

[107] Zhu F, Li T, Zhang L, Liu N. Co-up-regulation of three P450 genes in response to permethrin exposure in permethrin resistant *house flies, Musca domestica*. BMC Physiology. 2008;**8**(1):1-13

[108] Zhang G, Zhang W. Protein–protein interaction network analysis of insecticide resistance molecular mechanism in *Drosophila melanogaster*. Archives of Insect Biochemistry and Physiology. 2019;**100(1)**:e21523

Section 3

Insect Pest Control by Pheromones Plant Extract

Chapter 6

Potency of Neem, *Azadirachta indica* L. (A. Juss) Leaf Aqueous Extract Insecticide against White Mango Scale, *Aulacuspis tubercularis* Newstead (Homoptera: Diaspididae) Infesting Mango (*Mangifera indica* L.)

Temesgen Fita, Emana Getu, Mulatu Wakgari and Kebede Woldetsadike

Abstract

This study was conducted to evaluate the potency of aqueous *A. indica* leaf powder extract (LPWE) against *A. tubercularis* infesting mango. Field experiments were conducted in Ethiopia at two experimental sites, western Oromia. Randomized Complete Block design was laid in four replications with four blocks consisting of sixteen treatment plots. The treatments were 0.05, 0.1 and 0.15 mg/ml spray concentrations made from *A. indica* LPWE. The treatments were applied 3 times at 10-day interval and the mortality count was carried out 10 days after 1st, 2nd and 3rd treatment applications. The results of the study confirmed that among the tested three different concentration of *A. indica* LPWE 0.15 mg/ml concentration significantly ($p < 0.05$) reduced the population of sessile *A. tubercularis* at both experimental sites. The results also indicated that male adults and nymphs were more affected than females. The population of sessile *A. tubercularis* significantly decreased as the concentration of *A. indica* LPWE increased in comparison with the check plots. Thus, the use of *A. indica* LPWE at high concentrations led to a notable population reduction of sessile *A. tubercularis* and its effects. Hence, the use of *A. indica* LPWE can be recommended for *A. tubercularis* management as part of integrated pest management.

Keywords: *Azadirachta indica*, aqueous extract, *Aulacaspis tubercularis*, botanical insecticide, management, nymphs, potency

1. Introduction

Mango is one of the most widely cultivated fruit crops in Ethiopia, preceded only by bananas in terms of economic importance. Most of the mango productions come mainly from the southwestern, western, central rift valley, and eastern parts of Ethiopia [1] The annual mango production in Ethiopia is 151,331.24 t with an area coverage of 20,783.92 ha, and its production is 7.28 tones ha^{-1} [2] which accounts for about 0.18% of the world production. Mango is attacked by multiple arthropod insect pests among the damage inflicted by the white mango scale, and *Aulacaspis tubercularis/mangiferae* Newstead (Homoptera: Coccoidea: Diaspididae) is the most important insect pest of mango [3, 4].

This insect is a cosmopolitan pest throughout the globe regardless of where mango is produced [5, 6]. *A. tubercularis* is a tropical species believed to have originated from tropical Asia [7]. White mango scale *A. tubercularis* global dispersal could have taken place through the movement of plant materials [8]. This insect pest was first identified in Ethiopia in 2010 invading private commercial mango plants in East Wollega Zone, Oromia Regional State, western Ethiopia [3] from where it was dispersed to distinct mango-growing parts of the country [4]. At the climax level of infestations, *A. tubercularis* generates production-dropping covering from 50 to 100% [9].

Utilization of the synthetic chemical insecticide to control insect pests causes many environmental problems such as destabilizing the biodiversity and trophic level of the ecosystem, harmful effects on animals, human health, and the beneficial insects [10]. It has also been observed that though the chemical industry is aware of the environmental effect of the misuse of pesticides, they are not giving due regards to promotion of ecologically sound practices that will enhance sustainability in agricultural production [11]. Reducing chemical pesticide use has become globally shared issues in several countries and become a major issue in public policies due its negative impacts on the environment and on human health [12]. Homemade botanical insecticides are widely used by subsistence and transitional farmers in low-income countries. Their use is often driven by the limited availability or cost of commercial pesticides. Homemade botanical insecticides are often recommended by agricultural extension services and some development organizations. However, this could be questioned because scientific evidence of their efficacy and safety may not be available or accessible [13].

One of the most economically important potential plants to become a vegetative pesticide is the neem tree, *Axadirachta indica L.* (A. Juss.), which contains azadirachtin that is accumulated on the leaves and specifically in the seeds [14]. Neem-based pesticides play a vital role in pest management and hence have been widely used in agriculture. Mostly all parts of the neem tree like the leaves, bark, flowers, seeds, and fruit pulp are used in the powdery or in the extract form like leaf extract, kernel extract, cake extract, oil spray, etc. [15]. Neem *A. indica* is a plant having evergreen leaves all year round; however, the fruits and seeds can only be available once a year. In recent times and following the isolation of azadirachtin, the major active compound, that is mainly responsible for the insecticidal activity of neem, the use of neem-based insecticide has increased in the last 30 years, and it is currently the most successful botanical pesticide in agricultural use worldwide [16, 17]. Azadirachtin acts as an antifeedant, repellent, and repugnant agent and induces sterility in insects by preventing oviposition and interrupting sperm production in males [16, 18, 19]. When comparing neem oil formulations with synthetic larvicides, it is costly but neem oil was more effective than the synthetic one for preventing pest resistance [20]. *A. indica* seed powder water extracts at 0.15 concentration have a better impact on knocking down the population of *A. tubercularis*, and it can potentially be used for the management of the newly emerging and inflicting mango pest, *A. tubercularis* [9]. Extract of neem, *A. indica*, works in a way of interfering with the reproduction, growth of insect, and behavior in the forms of repellent, attractant, antifeedant, and poisoning larva and imago, either as a pesticide or as a contact poison [21]. However, its mechanisms of action still unclear and remain to be clarified, especially in relation to the neurophysiological and the possible long-term activities [22]. It is eco-friendly and has nontoxic actions, and the peculiar mode of action as well as its broad-spectrum activity over chemical pesticides has offered many advantages to beneficial organisms. Like to *A. indica* seeds, *Azadirchatin* is also present in *A. indica* leaf with moderate concentrations, and its availability all-round the year make to give focus to the leaf. Therefore, this research aimed on assessing the potency of *A. indica* leaf powder water extract (LPWE) against *A. tubercularis* as a contribution to the control of this mango damaging insect.

2. Materials and methods

2.1 Study area description

The insecticidal potency test of neem *A. indica* leaf powder extract was carried out during the cropping season of March to May 2018 at Uke Kersa Farmers Administrative of Guto Gida district and Arjo Gudatu Farmers Administrative Kebele, Digga District of East Wollega Administrative zone of Oromia Regional State, western Ethiopia. The selected two study areas represent mango-growing areas of western Oromia Regional State. The ecological zones of the study areas are lowlands (wet kolla), (**Table 1**) with slight altitudinal differences, and both sites are suitable for mango production (**Figure 1**).

2.2 Research materials preparation

Deep green leaves of neem trees piled up from Dire Dawa city administration, east Ethiopia. The collected leaves were cleaned exhaustively by clean water at its crisp stage to avoid any waste matter or any accumulations amassing materials from the collected leaves, which may reduce the potency of the final *A. indica* leaf extract hurt the apparatus during rectification. The then under shade, plastic sheets were placed on wooden benches and the cleaned leaves were thoroughly distributed *on the sheets* for good air circulation until absolute green drying. The dried leaves were squashed and stocked in cloth sack. In accordance with the [10, 19] procedures, the crushed *A. indica* neem leaves were ground gently to make fine powder using coffee grinder (Coffee and Spice Grinder 220-240 V 50-60HZ', model no. SZJ-830 'S SAYONA Patirrier DELUXE). The ground *A. indica* leave screened by 1 mm^2 mesh wire to obtain a fine particle. Obeying the [23] procedures, with some adjustment, the readymade leaf powder was measured at an amount of 0, 50, 100, and 150 mg and was added into four plastic bucket each containing 15 l of pure water to make up a 0, 5, 10, and 15% solvent, respectively. Then the neem leaf powder extracts in each bucket were mixed gently and vortexed very well. The readymade solvents were stored in the laboratory at an ambient temperature of 27 ± 2°C throughout the period of the study. The leaf extract was filtered by using a muslin bag cloth. To stick the pure filtrate of neem powder water extract on

Districts	Coordinates		Altitude	Mean temp. (0 °C)			RH (%)	Rainfall (mm)
	N	E		Min.	Max.	Aver.		
G.G	9°19′ 01.86	36° 30′42.22”	1391	9.93	37.12	27.2	69.0	1725.98
Diga	9°02.225″	036° 15.013″	1304	8.24	33.37	25.72	74.1	2089.89

NB: G.G. = Guto Gida; Min. =minimum; Max. =maximum; Aver. = average; m.a.s.l. = meter above sea level; mm = millimeter; RH = relative humidity.

Table 1.
Argo-ecological and coordinates of the study area [Source: Ethiopian Meteorology Agency (EMA)].

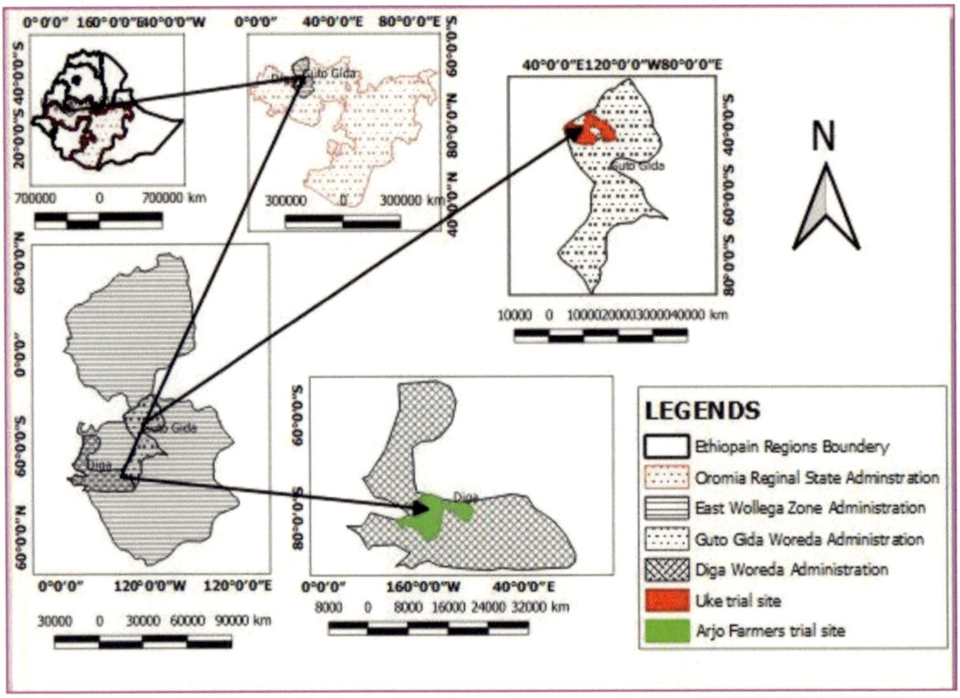

Figure 1.
The log concentration and probit mortality regression line graph for A. indica *LPWE treatments against sessile WMS (nymphs, adult females and males) after 1st, 2nd, and 3rd round treatment at Arjo farmers' association administration experimental site.*

the leaf surface of the mango plants, soap with no detergent was added at the rate of 1 ml/liter as an emulsifier [24].

2.3 Study design and treatment application

The experiments were conducted during the cropping season of April to May 2018 at two experimental sites of Arjo Gudetu and Uke Kersa farmers administration of Diga and Guto Gida districts, east Wollega zone of Oromia Regional State, western Ethiopia. Mango plants of similar size and age were considered for the experiment. For the last 2 years, no pesticides were applied to the mango trees. Randomized complete block design (RCBD) was laid in four replications with four blocks consisting of 16 mango plants as treatment plots. A plot consists of one mango plant. Spacing between plants and rows was 7 m and 8 m, respectively. The block size was $224m^2$ having four mango plants. Separate three treatment concentrations were made ready from neem leaf powder extracts. The treatments were 0.05, 0.1, and 0.15 mg/ml concentration, and the check plots (neutral water) were used as a negative control for comparison. The *A. indica* aqueous leaf powder extracts were sprayed in four replications.

The concentrations were calculated using the following formula (Eq. (1)):

$$C_1 V_1 = C_2 V_2, \quad (1)$$

where C_1 and C_2 represent initial and final concentration, respectively, and V_1 and V_2 represent initial and final volume, respectively [25].

After the *A. tubercularis* population approximately reached the economic injury level (EIL) (above five clusters population per leaf), the application was commenced and continued for three rounds at an interval of 10 days. A randomized complete block design (RCBD) was laid out for a field experiment in four replications. The mortality count of sessile *A. tubercularis* was recorded at after 7 days of the 1st, 2nd, and 3rd treatments application. A manually calibrated 'Knapsack Sprayer (Jacto16 HD400 Sprayer Thailand made) with a capacity of 16 l was used for treatment application. Accordingly, for complete spray, 12 l of water was used for each plant. Even though there is no established economic threshold level (ETL) of *A. tubercularis*, assuming that when the pest population reached the economic injury level (EIL), treatment application was commenced and repeated every 10 days from 20 April to 20 May 2018 for three rounds. Spraying of the neem leaf extract was carried out in the afternoon at 3:30 pm to reduce the loss of the chemicals due to evaporation.

2.4 Data collection

Twelve leaves were plucked from the sprayed mango trees from the middle, lower, and top canopies at four cardinals of each sampled tree 10 days after treatment application (DATA). The collected samples were reserved in paper bags and labeled accordingly. Samples were then put into plastic bugs, transported within a day to the School of Veterinary Science Laboratory, Wallaga University, and reserved at room temperature of 27 ± 2°C. On the next day, the sampled mango leaves and the number of deadly nymphs and adults of *A. tubercularis* were counted under a dissecting stereomicroscope. The reduction percentage of the *A. tubercularis* population was the basis of the evaluation of the tested botanical aqueous extract. The counted dead *A. tubercularis* insect nymphs and adult data were converted into corrected percent mortality. Pre- and post-spray counts of the nymphs and adults per leaf were also recorded from the sampled leaves, and the decrement in nymph and adult numbers by the application of aqueous *A. indica* LPWE (efficacy %) was computed based on the [26] equation (Eq. (2)).

$$\text{Efficacy (corrected)\%} = \left(1 - \frac{\text{n in Co before treatment} * \text{n in T after treatment}}{\text{n in Co after treatment} * \text{n in T before treatment}}\right) * 100 \quad (2)$$

where n = Insect population, T = treated, and Co = control.

Due to probable phytotoxicity, any change in the color and texture of mango leaves treated with neem leaf powder extract was noted. The outcome of the deadly nymphs and adults count were expressed in percentage (D.P. %) with correction factor for control/check plots using Abbott's formula [27] as (Eq. (3)).

$$\text{Mortality}\,(\text{corrected})\% = \left(1 - \frac{\text{n in T after treatment}}{\text{n in Co after treatment}}\right) * 100 \qquad (3)$$

where n = Insect population, T = treated, and Co = control.

Nymph and adult stages of *A. tubercularis* were regarded as extinct/lifeless if there is alter in color (darkened or dusky), desiccated and or vacant, and motion-less of organs/protuberance when stroked/patted with a feathery brush.

2.5 Statistical analysis

The collected data were subjected to Statistical Analysis System (SAS) software version 9.0 [28]. The mixed procedure repeated-measure with restricted maximum likelihood (REML) models were employed for statistical analysis of variation between experimental units [29]. Tukey's honestly significant difference (HSD) method was used for mean separation at P < 0.05 level of significance. The LC_{50} and LC_{95} of the treatment were calculated by Finney's probit *data* analysis method *using an* MS Excel worksheet [30]. The confidence limits deployed on the treatment concentrations and the probit-mortality of *A. tubercularis* resolved by the logistic regression method [31].

3. Results

3.1 Potency of *A. indica* LPWE against sessile *A. tubercularis* nymphs and adults

The results of the study revealed that mortality of *A. tubercularis* nymphs and adults raised significantly (p < 0.001) with an increase in the concentration of *A. indica* leaf powder extract. The mortality of nymphs and adults of *A. tubercularis* due to different concentrations of *A. indica* LPWE at Uke Kersa and Arjo Farmers' Administrative Kebele experimental sites is shown in **Table 2**. At both experimental sites, the highest mean percent mortality rate of sessile *A. tubercularis* (nymphs, adult males and females) were (±SE) 72.28 and 73.16 at Uke Kersa and Arjo Farmers experimental sites, respectively, during the 3rd round treatment applications with 15% aqueous *A. indica* LPWE treatment concentrations. Moreover, the highest mean percent mortality was recorded on nymphs with 15% *A. indica* LPWE during the 3rd round of treatment with mortality of 87.18 and 89.06 at Uke Kersa and Arjo Farmers experimental sites, respectively. However, the least mean percent mortality was scored with a 5% treatment concentration, during the 1st round of treatment application with mortality of 30.81 and 32.58 at Uke Kersa and Arjo Farmers experimental sites, respectively. Under control treatments, natural deaths were also observed due to hot environmental conditions.

The result of the study depicted that increased mortality of sessile *A. tubercularis* increased with an increase in the concentration of aqueous *A. indica* LPWE

A. indica LPWE concentration %	Round of TA	DATA	Mean (±SE) percent mortality	
			Arjo Gudetu	Uke Kersa
0 (control)	1st	10	15.12 ± 0.22a	14.78 ± 0.64a
	2nd	20	20.73 ± 0.38a	19.11 ± 0.61a
	3rd	30	20.94 ± 0.44a	18.56 ± 0.80a
Mean			18.93 ± 0.23a	17.61 ± 0.68a
5	1st	10	30.81 ± 0.22b	32.58 ± 0.64b
	2nd	20	49.28 ± 0.22c	45.85 ± 0.64c
	3rd	30	60.52 ± 0.22 cd	52.08 ± 0.60c
Mean			50.93 ± 0.23c	53.95 ± 0.68c
10	1st	10	50.89 ± 0.38c	55.75 ± 0.61c
	2nd	20	63.31 ± 0.38 cd	66.77 ± 0.61 cd
	3rd	30	71.79 ± 0.38d	75.31 ± 0.61d
Mean			64.56 ± 0.23 cd	65.48 ± 0.68 cd
15	1st	10	71.08 ± 0.44d	73.12 ± 0.80d
	2nd	20	81.08 ± 0.44e	83.43 ± 0.80e
	3rd	30	87.18 ± 0.44e	89.06 ± 0.80e
Mean			73.16 ± 0.23d	72.28 ± 0.66d
Df			28	28
F value			52.20	48.48
Pr > f			<.0001	<.0001

* *Within a column means go along with alike letter (s) are not notably unlike from each other (P < 0.05) by Tukey's studentized range test (HSD).*

Table 2.
The potency of A. indica *LPWE treatment application on* A. tubercularis *nymphs and adults during consecutive three-round treatment application (TA) and days after treatment application (DATA) at Arjo Gudetu and Uke Kersa farmers experimental sites.*

treatments application. Female *A. tubercularis* were more tolerant than males and nymphs to all levels of aqueous *A. indica* LPWE treatments concentration. From the result of the experiment, it was clearly observed that the 1st-round treatments application caused insignificant mortality as compared to the 2nd- and 3rd-round treatments application. It is because of that the scale insects' body is covered with external protective scale layers in which a single spray cannot penetrate/reach the scale insects unless otherwise a redundant spray could be applied. At both experimental sites, the lowest mean percent mortality of *A. tubercularis* was recorded at the first round (10 days after) treatment application, while the highest was recorded after the 3rd round (30^{the} day) of treatment applications.

Aulacaspis tubercularis nymphs and adults mean percent mortality due to application NLPWE at Uke Kersa and Arjo Gudetu Farmers Association are shown in **Table 3**. The minimum mean fatality percent was observed in females followed by males and crawlers in that order.

A. indica LPWE concentration %	Sex and stages of *A. tubercularis*	Mean (±SE) percent mortality	
		Arjo Gudetu	Uke Kersa
5	Male	43.11 ± 0.34d	40.63 ± 0.53d
	Female	21.97 ± 0.34f	19.36 ± 0.53f
	Nymphs	51.72 ± 0.34c	48.98 ± 0.51c
10	Male	55.34 ± 0.50b	57.48 ± 0.77b
	Female	35.46 ± 0.50e	37.27 ± 0.77e
	Nymphs	64.25 ± 0.50ab	67.95 ± 0.78ab
15	Male	69.58 ± 0.52a	70.06 ± 0.83a
	Female	48.31 ± 0.52b	51.09 ± 0.83b
	Nymphs	77.33 ± 0.52a	76.98 ± 0.83a
Df		28	28
F value		52.20	48.48
Pr > f		<.0001	<.0001

** Within a column means go along with alike letter (s) are not notably different from each other (P < 0.05) by Tukey's studentized range test (HSD).*

Table 3.
Insecticidal activity of A. indica *LPWE treatment application on sessile white mango scale nymphs and adults at different stages and sexes due to different concentrations of treatments application at Uke Kersa and Arjo Gudetu farmers' experimental sites.*

Site name	DATA	LC50 (μg/ml)	LL-UL	LC95 (μg/ml)	LL-UL	Slope ± SE	(X^2)
AG	10	15.4	13.98–1687	153.9	102.1–231-8	4.66 ± 0.98	118.0b
	20	8.9	8.25–9.7	174.9	103.6–295.1	5.01 ± 0.96	155.4b
	30	5.1	4.49–5.7	79.5	54.8–115.4	5.33 ± 0.95	212.7 cd
Uke	10	28.2	22.0–36.0	752.5	298.7–1895.9	4.46 ± 0.98	94.8a
	20	11.4	10.84–12.1	67.3	53.8–83.99	4.82 ± 0.97	65.6a
	30	4.9	4.49–5.34	34.2	28.5–41.03	5.46 ± 0.92	267.5c

DATA = days after treatment application; AG = Arjo Gudetu.

Table 4.
LC50 and LC95 of A. indica *LPWE application against* A. tubercularis *days after treatment (DAT) application at Arjo and Uke Kersa farmers' association administration experimental sites.*

The contact toxicity of *A. indica* LPWE against white mango scale nymphs and adults of *A. tubercularis* is shown in **Table 4**. The LC_{50} values of *A. indica* LPWE at Arjo Farmers Association administration experimental site against sessile nymphs and adults of *A. tubercularis* during 1st, 2nd, and 3rd round treatments application were 15.4, 8.9, and 5.1 mg/100 ml, respectively, while the Chi-square (X^2) values with 1st, 2nd, and 3rd round treatments application were 118.0, 155.4, and 212.7, respectively. The LC_{50} values of aqueous *A. indica* leaf powder at the Uke site against sessile *A. tubercularis* at 1st, 2nd, and 3rd round treatment application were 28.2, 11.4, and 4.90 mg/100 ml, respectively, while the Chi-square (X^2) for 1st, 2nd, and 3rd round treatments application were 94.8, 65.6, and 267.5, respectively.

At Arjo Farmers experimental site, the contact toxicity for essential extracts of aqueous *A. indica* LPWE at 1st round treatment application against different stages of *A. tubercularis* adult nymphs, females and males had the LC_{50} values of 10.6, 24.5 and 14.0 μg/ml, respectively.

The contact toxicity of aqueous *A. indica* leaf powder at 2nd round treatments application against adult nymphs, females, and males had LC_{50} values of 5.5, 18.3, and 7.9 μg/ml and for the 3rd round treatments application had LC_{50} values of 3.5, 10.2, and 4.5 μg/ml, respectively. The sessile nymphs and adults of *A. tubercularis* fatality rate count were made consequently 10 days after each round of treatment application. With the same activity at Uke Kersa farmers' association administration experimental site, the essential extracts of *A. indica* LPWE application for 1st round treatments against different stages of *A. tubercularis* adult nymphs, females, and males showed LC_{50} values of 17.8, 49.7, and 24.7 μg/ml, respectively. The contact toxicity of *A. indica* LPWE for 2nd round treatments application against *A. tubercularis* nymphs, adult females, and males showed LC_{50} values of 9.9, 17.6, and 11.1 μg/ml, respectively. With the same treatments of *A. indica* LPWE, at 3rd round application against *A. tubercularis* of nymphs, adult females and males showed

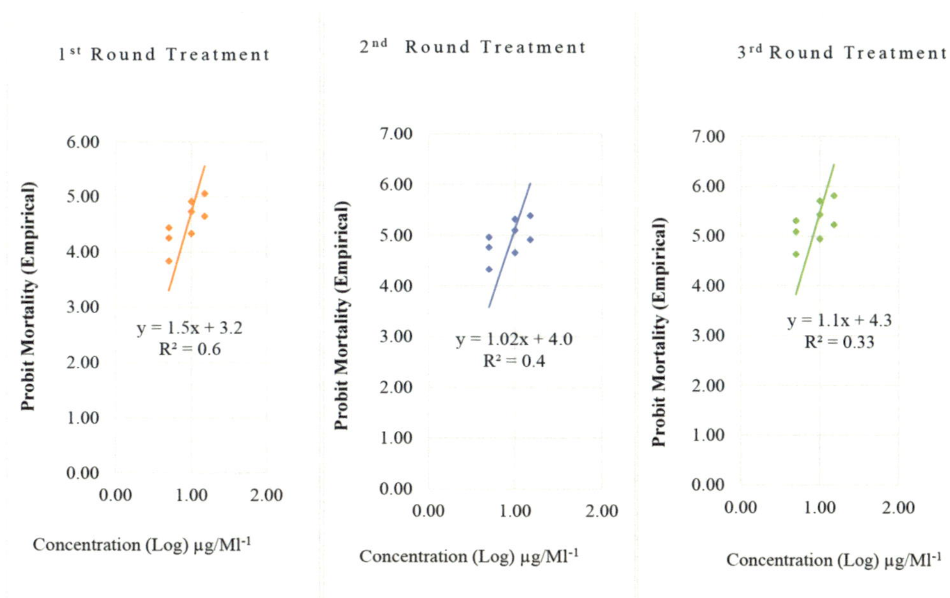

Figure 2.
*Study area map for evaluation of neem (*Azadirachta indica*) leaf powder potency against* Aulacaspis tubercularis *in Oromia regional state, Western Ethiopia.*

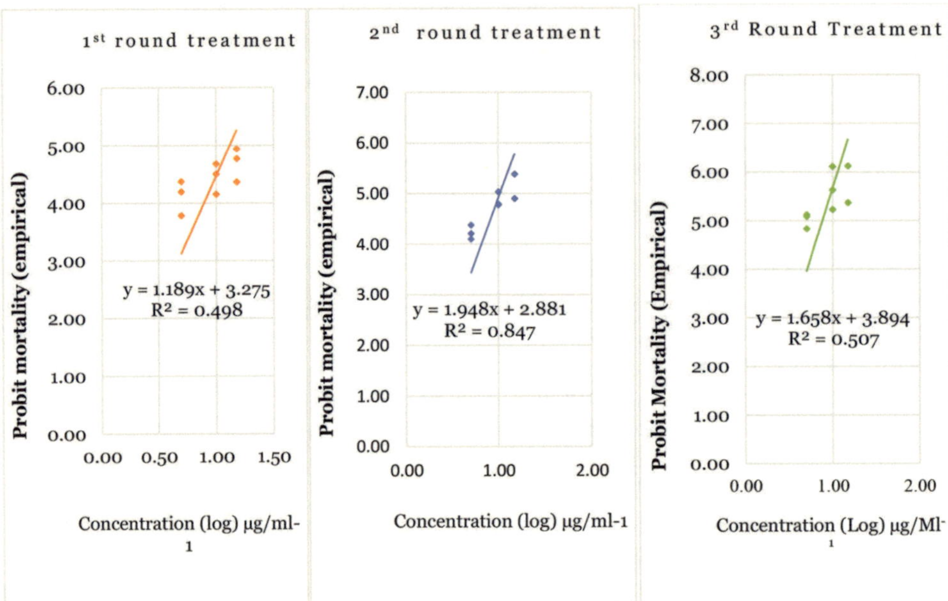

Figure 3.
The log concentration and probit mortality regression line graph for A. indica *LPWE treatments against sessile WMS (nymphs, adult females and males) after 1st, 2nd, and 3rd round treatment at Uke Kersa farmers' association administration experimental site.*

LC_{50} values of 4.8, 6.8, and 4.4 µg/ml, respectively. At both experimental sites, the Chi-square values were significant at a $P \leq 0.05$ level at the treatments were promising for the management of sessile *A. tubercularis.*

At Arjo Farmers Association administration treatment site, the log-probit regression line calculated (**Figure 2**) values were Y = 1.5x + 3.2 (X-0.6), Y = 1.0x + 4.0 (X-0.4), and Y = 1.1x + 4.3 (X-0.3) for 1st, 2nd, and 3[d] round treatments application, respectively. Likewise, at Uke experimental site (**Figure 3**) the log-probit regression line calculated values were Y = 1.2x + 3.3 (X-0.5), Y = 1.9x + 2.9 (X-0.8), and Y = 1.7X + 3.9 (X-0.5) for 1st, 2nd, and 3rd round treatments, respectively. At both experimental sites, the R^2 values of the three-round treatment application showed significant differences (**Figures 2** and **3**). The result reviled that the regression model analysis showed notable differences that indicate the mortality caused by the tested three *A. indica* LPWE concentrations showed notably different outputs, designated that the fatality brought about during the 3rd cycle treatment application (30 DAT) showed progressive output than the 1st and 2nd cycle treatment applications. The essential extracts of *A. indica* LPWE application at Arjo and Uke Kersa farmers association administration experimental sites against sessile *A. tubercularis* adult females, males, and crawlers are shown in **Table 5**. The increment in the likelihood of mortality rate was commensurately significant at lower log treatment concentrations than the higher ones, so the regression bar was raised at the lower tips than at the higher tips.

In both study sites, there was statistically significant ($p < .01$) positive correlation between aqueous *A. indica* LPWE, location, and days after treatment (DAT) with *sex* and mortality of white mango scale populations as depicted in **Table 6**. Moreover, there was a slight negative correlation between both locations (experimental sites)

Study sites	Stage of WMS	DAT	LC_{50} µg/ml	LL-UL	LC_{95} µg/ml	LL-UL	Slope ± SE	(X^2)
AF	*Nymphs*	10	10.6	9.6–11.8	92.0	54.7–158.3	4.89 ± 0.97	0.54bc
	Female	10	24.5	18.7–32.0	231.7	98.6–554.5	4.34 ± 0.99	0.01a
	Male	10	14.0	12.2–16.2	132.6	69.6–252.9	4.72 ± 0.98	0.16b
	Nymphs	20	5.5	4.6–6.6	78.7	42.8–144.8	5.29 ± 0.95	0.94d
	Female	20	18.3	14.0–23.9	422.3	118.0–1516.0	4.65 ± 0.97	0.27b
	Male	20	7.9	6.8–9.1	150.1	64.0–352.5	5.08 ± 0.96	0.45bc
	Nymphs	30	3.5	2.8–4.4	29.5	21.5–40.5	5.67 ± 0.92	3.45f
	Female	30	10.2	8.8–11.8	225.6	81.1–627.5	4.94 ± 0.96	0.69c
	Male	30	4.5	3.6–5.6	63.2	36.1–110.4	5.42 ± 0.94	1.54e
UK	*Nymphs*	10	17.8	13.6–23.2	464.8	119.9–1802.4	4.68 ± 0.97	0.40bc
	Female	10	49.7	25.7–96.2	1135.5	173.8–7420.4	4.15 ± 0.99	0.00a
	Male	10	24.7	17.1–35.9	613.7	139.3–2704.1	4.51 ± 0.98	0.51c
	Nymphs	20	9.9	9.3–10.5	35.0	28.8–42.5	4.94 ± 0.98	0.18b
	Female	20	17.6	13.3–23.3	563.5	124.4–2551.8	4.70 ± 0.97	0.73 cd
	Male	20	11.1	10.3–12.0	54.1	39.6–73.9	4.84 ± 0.98	3.59f
	Nymphs	30	4.8	4.4–5.2	13.1	11.8–14.4	5.67 ± 0.86	2.61ef
	Female	30	6.8	5.7–8.2	191.1	67.5–541.2	5.15 ± 0.95	0.48c
	Male	30	4.4	3.8–5.2	29.8	22.6–39.3	5.56 ± 0.93	1.90e

**AF = Arjo Farmers; UK = Uke Kersa; WMS = white mango scale; DAT = days after treatment; LC_{50} and LC_{95} values are expressed as percentage; LC_{50} = median lethal concentration.*

Table 5.
Different concentration of A. indica *LPWE applications for LC50 and LC95 against* A. tubercularis *nymphs, adult females and males (n = 360) 10th, 20th, and 30th DATA application at Arjo farmers and Uke Kersa experimental sites.*

		Pearson's correlation coefficient (r) and p (r)				
		Location	Treatments	Sex	DATA	Mortality %
Location	r	1				
	P(r)					
Treatments	r	0.00	1			
	P(r)	1.00				
Sex	r	0.001	0.001	1		
	P(r)	0.962	0.949			
DATA	r	0.00	0.000	0.000	1	
	P(r)	1.00	1.000	1.000		
Mortality %	r	−0.004	0.702**	0.126**	0.408**	1
	P(r)	0.834	0.000	0.000	0.000	

***. Correlation is significant at 0.01 level (2-tailed).*

Table 6.
Correlations of location, A. indica *LPWE treatments and days after treatment application (DATA) with sex and mortality of* A. tubercularis *populations in Uke and Arjo farmers orchards.*

and mortality of *A. tubercularis* nymphs, adult females and males with *A. indica* LPWE treatments application, which could be due to slight agro-ecological differences of the study sites with some basic climatic factors such as temperature, relative humidity, and rainfall.

4. Discussion

The current study illustrated the potency level of *A. indica* LPWE for the management of the white mango scale, *A. tubercularis.* The study by [32] on neem-borne molecules as eco-friendly control tools against mosquito vectors of economic importance revealed that neem *A. indica* extracts are an eco-friendly pest control tool, which has an attractive crown of deep-green foliage available throughout the year. The present study revealed that the application of *A. indica* LPWE is helpful for the management control of sessile *A. tubercularis*. At both study sites, the mortality percentage of *A. tubercularis* with the use of *A. indica* LPWE showed moderately a slightly different mortality result, suggesting that there are some basic climatic factors variations. Nonetheless, there were notable variations observed in mortality between nymphs, adult females, and males of *A. tubercularis* because of variations in their vulnerability to the administration of *A. indica* LPWE. Moreover, adult females' have more tolerance in terms of relatively less mortality than nymphs and male against *A. indica* LPWE at Arjo and Uke Kersa experimental sites, signifying that the growth stages and sex of the insect adversely responded to the application of the *A. indica* LPWE treatments. A study by [33] stated that the mixture of *A. indica* seed oil (Trilogy) against the white mango scale was effectual which caused 76.92 and 81.03% fatality in adults (females and males) and the nymphs, respectively. In contrast, the present study showed significant difference among mortality of male, female, and nymphs at different concentrations of *A. indica* LPWE at both experimental setups. A study by [9] also mentioned that the application of *A. indica* seed powder water extract (SPWE) treatments caused mean mortality of 86 and 83% in sessile *A. tubercularis* at Arjo Gudetu and Uke Kersa experimental sites of east Wollega zone of Oromiya Regional State, western Ethiopia. The high Chi-square values in the treatments probably indicated the heterogeneity of the tested populations. Different concentration levels of *A. indica* LPWE influenced the mortality of sessile *A. tubercularis* (nymphs, adult females, and males) differently. At both experimental sites, the control plot (distilled water) did not show significant mortality rather than natural death.

The outcome of the current study indicated that the *A. indica* LPWE at a 5% concentration against the white mango scale can be utilized as a protective implementation to reduce initial infestation. Furthermore, the result of our study indicated that accurate administration of *A. indica* LPWE including the underside leaf surfaces resulted in an effective management of the white mango scale.

Tolerance of *A. tubercularis* adult females to *A. indica* LPWE treatments could be because of the presence of wax cover or hard external scale coverage (exsuvial) of the insect, which is more impenetrable than the nymphs and males' wax cover. The result of our study goes with the result of [34], which mentioned that the muscular impenetrable wax covering that covers the body of the insect bear a protective obstruction against corporal and insecticide interference.

The finding of the present study also indicated that the botanical formulations from *A. indica* LPWE could put back the use of synthetic insecticides in Integrated Pest Management (IPM) programs. The finding of this study goes in line with the result of [33] who mentioned that neem derivative botanicals have an insecticidal effect for scale insect control and are environmentally friendly and useful in reducing the environmental pollution. He also stated that in the present situation, the neem (*A. indica*) derivative is a vigorous botanical insecticide selected for organic agriculture that is widely used in many countries around the world either singly or in IPM or in coexistence with commercial pesticides. The findings of the present study is supported by [35] results, which mentioned that neem derivative compounds are an eco-friendly botanical insecticide against armord scale insects and mealybugs, and it has a considerable reaction as an insecticide on the population reduction of white mango scale *(Aulacaspis mangiferae)*. The study also revealed that *A. indica* crucial oils can be used likely as a substitute source for advancing bio-insecticides development against scale insects.

This showed that the resident community can use *A. indica* LPWE to manage the invasion of *A. tubercularis* in plantation mangoes. Accordingly, the output of this study revealed that the 15% concentrations of *A. indica* LPWE at three or more sequences/frequency of application has a better efficacy on population depletion, and likely it can be utilized for the control of the white mango scale, *A. tubercularis.*

5. Conclusion

The insecticidal properties of natural plant products have been known since the ancient times. Among the various plant products used as insecticides, the formulations developed from neem (*A. indica*) have shown promising result for insect pest management. The botanical formulations derived from *A. indica* are a nontoxic, biodegradable, and environmentally friendly. In this regard, considering the high risks of chemical insecticides on environmental safety, natural enemies, and animals as well as the human beings, the botanical extract are valuable, cheap, safe, and environmentally friendly alternative insect pest management options. In Ethiopia, considering the management options toward *A. tubercularis* control is almost nil or little, which enabled the pest to invade the whole mango-growing parts of the country. Thus, the result of this study presented an encouraging outcome that *A. indica* aqueous LPWE at 15% application has a promising result to knock down *A. tubercularis* population. Therefore, the aqueous *A. indica* LPWE as a derivative botanical pesticide, besides to implementation of cultural management practices of *A. tubercularis,* such as consistent scouting and periodic inventory systems as well as periodic pruning for *A. tubercularis* infestation, removal, and burning of the infested branches, are important practices as part of IPM.

Thus, the development and expansion of the neem (*A. indica*) tree in mango-producing belts of the country by incorporating into the "Green Legacy" program of the country could help for eco-friendly and nontoxic actions of pest control.

In addition, research actions on technological advancement in the utilization of neem insecticide should need exceptional consideration in the future plan of action with mango development programs.

Acknowledgements

We would like to thank HU, WU, and Ethiopian MoE for financial support partly in realizing this research work. Our appreciation also goes to Diga & Guto Gida Districts of Agriculture and Natural Resource Offices for their facilitation in availing experimental sites.

Funding

Haramaya University (HU), Wollega University (WU), and the Ministry of Education (MoE), made financial support partly for the research grant to the principal author.

Authors' contributions

Temesgen F. conceptualizes the study design, accomplishes the experiments and statistical analysis, and writes down the document. Prof. Emana G., Dr. Mulatu W., and Prof. Kebede Woldetsdik advise the principal author all over the work and evaluate the document. All coauthors look through and accepted the final document for publishing.

Conflict of interest

The research was conducted in the absence of any commercial or financial relationships that could be construed as a potential conflict of interest so that the authors declared that there has no conflict of interest.

Data availability statement

Primary raw data not presented in this study are available as supplementary material on request from the first author without undue reservation.

Author details

Temesgen Fita[1,2]*, Emana Getu[3], Mulatu Wakgari[1] and Kebede Woldetsadike[1]

1 School of Plant Sciences College of Agriculture and Natural Resources, Haramaya University, Ethiopia

2 Faculty of Agriculture and Natural Resources, Department of Plant Sciences, Wollega University, Ethiopia

3 Department of Zoological Sciences, Addis Ababa University, Ethiopia

*Address all correspondence to: temefita@gmail.com

References

[1] Neguse TB, Wanzala FKR, Ali WM, Owino WO, Mwangi GS. Mango (*Mangifera indica* L.) production practices and constraints in major production regions of Ethiopia. Academic Journals. African Journal of Agricultural Research. 2019;**14**(4):185-196

[2] CSA. The Federal Democratic Republic of Ethiopia central statistical agency. Agricultural Sample Survey Report on Area and Production of Major Crops. 2021;**2013 E.C**:1-143

[3] Fita T. White mango scale, *Aulacaspis tubercularis,* distribution and severity status in east and west Wollega zones, Western Ethiopia; Department of Plant Sciences, Wollega university. Science, Technology and Arts Research (STAR) Journal. 2014;**3**(3):01-10. DOI: 10.4314/star.v3i3.1. Available from: http://www.starjournal.org/uploads/starjournalnew/1(3).pdf

[4] Ofgaa Djirata, Emana Getu and R., Kahuthia-Gathu: A survey of geographical distribution and host range of white mango scale, *Aulacaspis tubercularis* Newstead (Homoptera: Diaspididae) in Western Ethiopia. Journal of Entomology and Nematology, 2019:11(5):59-65

[5] El-Metwally MM, Moussa SFM, Ghanim NM. Studies on the population fluctuations and distribution of the white mango scale insect, *Aulacaspis tubercularis* Newstead within the canopy of the mango trees in eastern of Delta region at the north of Egypt. Egyptian Academic Journal of Biological Sciences. 2011;**4**:123-130

[6] Ha S, Mahmoud Y, Ebadah I. Seasonal abundance, number of generations and associated injuries of the white mango scale, *Aulacaspis tubercularis (Mangifera)* (Newstead) (Homoptera: Diaspididae) attacking mango orchards. Research Journal of Pharmaceutical, Biological and Chemical Sciences. 2015;**6**:1373-1379

[7] Hamdy NM. Life table and morphometric studies of *Aulacaspis tubercularis* infesting mango trees in Egypt (Diaspididae: Hemiptera). Journal of Plant Protection and Pathology, Mansoura University. 2020;**11**(12):613-620. DOI: 10.21608/jppp.2020.135337

[8] Otieno MMO. A review of white mango scale (*Aulacaspis tubercularis* Newstead; Hemiptera: Diaspididae) in sub-Saharan Africa: Distribution, impact and management strategies. Pakistan Journal of Agricultural Research. 2021;**34**(1):227-238. DOI: 10.17582/journal.pjar/2021/34.1.227.238

[9] Fita T, Getu E, Wakgari M, Woldetsadike K. Efficacy of *Azadirachta indica* (A. Juss) seed powder water extract against *Aulacuspis tubercularis* Newstead (Homoptera: Diaspididae) on mango (*Mangifera indica* L.) In east Wollega, Ethiopia. AJOL, SINET: Ethiop. Journal of Science. 2020;**43**(1):11-20

[10] Pathak VM, Verma VK, Rawat BS, Kaur B, Babu N, Sharma A, et al. Current status of pesticide effects on environment, human health and its eco-friendly management as bioremediation: A comprehensive review. Fronteirs in Microbiology. 2022;**13**:1-29. DOI: 10.3389/fmicb.2022.962619

[11] Emmanuel O, Oludele OE, Unwabunne UM. Efficiency and economic viability of neem seed oil emulsion and cyper-diforce® insecticides in watermelon production

within the Nigerian southern Guinea Savanna zone. Journal of Plant Protection. 2019;**8**(1):81-101

[12] Jacquet F, Jeuffroy M-H, Jouan J, Le Cadre E, Litrico I, Malausa T, et al. Pesticide-free agriculture as a new paradigm for research. Agronomy for Sustainable Development. 2022;**42**:8. DOI: 10.1007/s13593-021-00742-8

[13] Dougoud J, Toepfer S, Bateman M, Jenner WH. Efficacy of homemade botanical insecticides based on traditional knowledge. A review. Agronomy for Sustainable Development. 2019;**39**(37):1-22. DOI: 10.1007/s13593-019-0583-1

[14] Ismael S, Ali M, Noe W, Djonwanwe MS, Eware BW, Cyrille. Evaluation of the efficiency of aqueous Extact of neem fruits on insect Pest of Rice in Rice agroecosystem of Maga in the far north region of Cameroon. International Journal of Environment, Agriculture and Biotechnology (IJEAB). 2018;**3**(5):1891-1905. DOI: 10.22161/ijeab/3.5.42

[15] Mondal E, Chakraborty K. *Azadirachta indica* - A tree with multifaceted applications: An overview. Journal of Pharmaceutical Science and Research. 2016;**8**(5):299-306

[16] Chaudhary S, Kanwar RK, Sehgal A, Cahill DM, Barrow CJ, Sehgal R, et al. Progress on *Azadirachta indica* based biopesticides in replacing synthetic toxic pesticides. Frontiers in Plant Science. 2017;**8**:610. DOI: 10.3389/fpls.2017.0061

[17] Pasquoto-Stigliani T, Campos EVR, Oliveira JL, Silva CMG, Bilesky Jose N, Guilger M, et al. Nanocapsules containing neem (*Azadirachta Indica*) oil: Development, characterization, and toxicity evaluation. Scientific Reports. 2017;**7**:5929. DOI: 10.1038/s41598-017-06092-4

[18] Aribi N, Denis B, Kilani-Morakchi S, Joly D. Lazadirachtine, un pesticide naturel aux effets multiples. Médecine Sciences. 2020;**36**:44-49. DOI: 10.1051/medsci/2019268

[19] Kilani-Morakchi S, Morakchi-Goudjil H, Sifi K. *Azadirachtin*-based insecticide: Overview, risk assessments, and future directions. Fronteirs Agronomy. 2021;**3**:676208. DOI: 10.3389/fagro.2021.676208

[20] Dua VK, Pandey AC, Raghavendra K, Gupta A, Sharma T, Dash A. P: Larvicidal activity of neem oil (*Azadirachta indica*) formulation against mosquitoes. Malaria Journal. 2009;**8**:124. DOI: 10.1186/1475-2875-8-124

[21] Suanda WI, Ni M. Delly Resiani: The activity of neem leaves, *Azadirachta indic*a (A. Juss.) extract insecticide as vegetative pesticide on Rice weevil (*Sitophilus Oryzae* L.) (Coleoptera: Curculionidae). Sustainable Environment Agricultural Science. 2020;**4**(1):10-17. DOI: 10.22225/seas.4.1.1520.10-17. Available from: http://ejournal.warmadewa.ac.id/index.php/seas

[22] Kumar SU, Abdulmadjid SN, Olaiya NG, Amirul AA, Rizal S, Rahman AA, et al. Extracted Compounds from Neem Leaves as Antimicrobial Agent on the Physico-Chemical Properties of Seaweed-Based Biopolymer Films. MDPI. Polymers. 2020;**12**(1119):1-17. DOI: 10.3390/polym12051119 . Available from: www.mdpi.com/journal/polymer

[23] Hurni H, Berhe WA, Chadhokar P, Daniel D, Gete Z, Grunder M, et al. Soil and Water Conservation in Ethiopia: Guidelines for Development Agents. Second ed. Bern, Switzerland: Centre for Development and Environment (CDE), University of Bern, with Bern Open Publishing (BOP); 2016. pp. 1-134

[24] Kwaifa NM, Ibrahim BI, Aliyu UA. Muhammad and Dangana A: Insecticidal effects of neem kernel extracts on flea beetle (Podagrica Uniforma J.) of okra (*Abelmoschus esculentus* L.) in Jega, Kebbi, Nigeria. IOSR Journal of Agriculture and Veterinary Science (IOSR-JAVS), Ver. II. 2015;**8**(12):57-60. Available from: www.iosrjournals.org

[25] Kudom AA, Mensah BA, Botchey MA. Aqueous neem extract versus neem powder on *Culex quinquefasciatus*: Implications for control in anthropogenic habitats. Journal of Insect Science. 2011;**11**:142. Available from: insectscience.org/11.142

[26] Henderson CF, Tilton E. W: Tests with acaricides against the brow wheat mite. Journal of Economic Entomology. 1955;**48**:157-161

[27] Abbott WA. A method of computing the effectiveness of an insecticide. Journal of Economic Entomology. 1925;**18**:265-267. DOI: 10.1093/jee/18.2.265a

[28] SAS (Statistical Analysis Software) Institute Inc. Moving and Accessing SAS® 9.2 Files. Cary, NC: SAS Institute, Cary, NC, USA; 2004. Available from: http://www.sas.com

[29] Smith AB, Diffey SM, Welsh AAH, B. R. Cullis: A new REML (parameter expanded) EM algorithm for linear mixed models. Australian and New Zealand Journal of Statistics. 2017;**59**(4):433-448

[30] Finney DJ. Probit Analysis: A Statistical Treatment of the Sigmoid Response Curve. London: Campridge University Press; 1947. p. 333

[31] Srinivasan MR, Kuttalam S, Chandrasekaran S, and Kennedy JS. Probit analysis: In Electronic Manual on Pesticides and Environment eds. Calculate LC_{50} or LD_{50} with MS Excel Worksheet Based on Finney's Method of Probit Analysis. Tamil Nadu Agricultural University, Coimbatore, 2017: DOI: 10.13140/RG.2.2.15957.42724 (assessed on Sep. 23, 2021)

[32] Nicoletti M, Murugan K, Canale A, Benelli G. Neem-borne molecules as eco-friendly control tools against mosquito vectors of economic importance. Current Organic Chemistry. 2016b;**20**(25):2681-2689. DOI: 10.2174/13 85272820666 1602 18233923

[33] Abdel-Aziz NF, Salem HA, Sammour EA, El-Bakry AM. Semi-field evaluation of some natural clean insecticides from essential oils on armored and soft scale insects (Homoptera: Diaspididae and Coccidae) infesting mango plants. National Research Centre, Dokki, Cairo, Egypt. International Journal of Chemical Technology Research. 2016;**9**(08):87-97

[34] Foldi I. The scale cover. In: Rosen D, editor. Armoured Scale Insects: Their Biology, Natural Enemies and Control. Amsterdam, 4A: Elsevier Science Publishers B.V; 1990. pp. 43-54

[35] Abdelaziz NF, Salem HA, E.A. Sammour: Insecticidal effect of certain ecofriendly compounds on some scale insects and mealybugs and their side effects on antioxidant enzymes of mango nurslings. Archives of Phytopathology and Plant Protection. 2014;**47**(1):1-14. DOI: 10.1080/03235 408.2013.800693

Chapter 7

Perspective Chapter: Current Situation of Insecticide Residues in Food Commodities and Possible Strategies for Management of Residues

Banka Kanda Kishore Reddy, Addanki Maneesha, Chinna Babu Naik, Malleswari Sadhineni, Tejaswi Yelleti and G. Raja Reddy

Abstract

Pesticides have evolved into a crucial instrument in agriculture's evolution as a plant protection agent for increasing food output. Moreover, pesticides contribute significantly by preventing a number of terrible diseases. However, both occupational and environmental pesticide exposure can lead to a number of health issues in people. It has been noted that pesticide exposures are becoming more and more associated with immune system suppression, hormone disruption, lowered intellect, abnormalities in reproduction, and cancer. Because of the great demand for farm produce and their lack of awareness of the hazardous consequences of pesticide residues in food, some farmers do not wait long enough for the residues to wash off after spraying before harvesting. As a result, residues in food products have appeared as a result of increased pesticide use in agriculture. Some of the primary tactics for reducing human exposure to pesticides are pesticide safety, regulation of pesticide usage, appropriate application technology, and integrated pest management.

Keywords: pesticide, pesticide residues, environmental pesticide exposure, pesticide safety, integrated pest management

1. Introduction

The usage of pesticides is common to ensure high agricultural yields. They are employed in the production of agricultural products as well as their post-harvest handling. The growing usage of chemical pesticides, however, has had a negative impact on human health as well as contaminated the environment. It has long been of great concern that food products may contain pesticide residues. When these products are eaten fresh, the issue becomes even more serious. Pesticides have been linked to a variety of risks to human health, from immediate effects like headaches and nausea to long-term ones like cancer, reproductive damage, and endocrine disruption. Pesticides applied during the fruit growth period can dissipate faster because of the growth dilution effect [1]. However, when applied after fruit growth, they are likely to be carried over to the harvested produce and processed products. Processing is considered as effective tool which implies transformation of fresh commodity into value-added product and ultimately affect the nature and magnitude of residues and during processing pesticide residues may increase or decrease in the transformed product [2]. Processing Factor (PF) is used to assess the risk associated with the consumption of pesticide residues, particularly for processed food products [3].

For pest management and the eradication of disease vectors, developing nations (like China) frequently utilize insecticides (including organophosphorous and pyrethroid) and fungicides (including triazoles and chloronitriles). Pesticide poisonings are far more common in poorer nations than in developed ones as a result of poor pesticide handling procedures, farmers' use of more harmful pesticides, and insufficient monitoring and oversight of these chemicals. Because of a lack of funding and the absence of strict regulations, pesticide residue control initiatives are frequently ineffective in poor nations.

1.1 Insecticide residues found in various food commodities

Insecticide residues discovered in various food commodities were tabulated in **Table 1**.

1.2 Management strategies to reduce pesticide residues

1.2.1 Effect of processing on the level of pesticide residues in various fruits and vegetables

Consumers have little control over pesticide residues that are left in food products in various proportions after harvesting and which are damaging to human health. Hence, pesticide residues present a significant challenge to the international trade in food products. Since pesticide-treated food crops always retain varying amounts of these chemicals, finding non-toxic methods for decontaminating food are essential. The molecular composition, product mix, and environmental conditions all have an impact

Name of the commodity	Insecticides found	Reference
Tomatoes	Oragno chlorines & Organo phosphates	[4]
Citrus	Chlorpyrifos, Deltamethrin and Spirotetramat	[5]
Tomato Ketchup	Cypermethrin	[6]
Wheat	Deltamethrin Permethrin Malathion	[7]
Rice	Diazinon & Chlorpyrifos	[8]
Apple	Abamectin, Diazinon & Chlorpyrifos	[9]
Sugar beet	Chlorpyrifos	[10]

Table 1.
List of insecticides found in various food commodities.

on the levels of pesticides in various food items. Washing is the most popular technique of processing, and it is an important first step in both home and industrial preparation. As a result, it is critical to consider techniques that may successfully assist in reducing residue content at the individual level in order to limit dietary pesticide exposure.

The effects of commonly used household processes such as washing by tap water, saltwater, lukewarm water, lemon water, tamarind water and ozone water are discussed below.

Tap water washing for 2 minutes eliminated 30–50% of phosalone residues and 65.3% of chlorpyrifos residues [11, 12], whereas tap water washing for 10 minutes removed 53.4, 53.3% dimethoate residues in grapes [13, 14]. Awasthi [15] found that washing mangoes with tap water removed 66–68% of the dimethoate and fenitrothion. Washing guava with tap water reduced dimethoate residues by 42.5–45.9% [16].

Other fruit crops viz., mangoes where 66% dimethoate residues were removed by tap water washing [15], 45.9% in guava [16]. Washing with salt water (2%) solution for 10 min was recorded as an effective decontaminant in removal of acephate, chlorpyrifos, quinalphos, bifenthrin residues (51.80–72.80%), acephate (72.74%), chlorpyrifos (67.52%) and quinalphos (65.0%), respectively in grapes [17–20], imidacloprid (61.89%) in field bean [21, 22], tetraniliprole (61.49%) in tomato [23] and NaCl (5%) removed 90% of quinalphos and profenofos in chili [24].

Pesticide residues combine with sodium chloride solution, a powerful electrolyte solution, which lowers their concentration and offers an appealing source for pesticide residue removal. When dipped in the solution, neonicotinoids with high water solubility were easily separated from the fruits in salt media. Vijayasree et al. [25] discovered that tamarind water (2%) and salt chloride (2%) solutions eliminated 85.56 and 100% of the emamectin benzoate in cowpea pods, respectively. Buprofezin residues in oranges were reduced by 36.50 and 27.51%, respectively, after washing with tap water and salt chloride solution (2%) [26].

Citric acid, the active component of lemon water, was mostly responsible for the residue elimination. The findings supported research in which a 52.2% reduction in dimethoate in tomato [16] was noted. Dimethoate and quinalphos were both eliminated by washing in lemon water (1%) for 10 minutes, along with 45% of the pesticide acetamiprid from tomatoes [27, 28].

The removal of residues by tamarind water solution is due to its acidic nature which is contributed by furan derivatives and carboxylic acids [29]. Studies where tamarind water (2%) washing resulted in a 69.1% reduction of chlorpyriphos in tomatoes, 58.8 and 80.4% reduction of dimethoate and quinalphos in grapes, 58.8% of dimethoate in grapes [14, 16, 27].

It is evident that the ozone concentration administered, the physical characteristics of the food matrix, and the residual ozone present in the medium all affect how effective ozone intervention for pesticide degradation is. The parameters that affect the clearance rate include the application environment (pH, temperature, and humidity), the application method (aqueous vs. gaseous), the ozone concentration, the rate of formation, and the geometry-size of pesticide residue [30]. Dipping apples in ozonated water of 0.25 ppm resulted in reducing the levels of azinophos-methyl on the surface of apples to 75% [31]. Ikeura et al., [32] studied the effect of ozone water (2.0 mg L-1) for 10 min on the level of fenitrothion residues in strawberries and removal rate were concluded as 25%. Removal of chlorpyrifos in lychee fruits with aqueous ozone water concentrations of 2.2, 2.4, 3.2 and 3.4 mg L-1 for 10, 20, 30 and 60 min resulted in 0, 25.8, 29.7 and 67.4% reductions, respectively. Similarly, fumigation of O_3 at 80, 160, 200 and 240 mg L-1 for 10, 20, 30 and 60 min resulted in 10, 18,

30, 45% reductions, respectively [33]. Treating the citrus fruits with ozonated water (10 mg L-1) for 5 min reduced the chlorpyrifos by 94.2% [34]. Washing of strawberries in ozonated water (1 mg L-1) for 5 min resulted in removal of chlorpyrifos by 71.5% [35]. Washing of tomatoes in ozonized water for 30 seconds removed chlorpyrifos by 86% [36].

To ensure that customers are not exposed to any health hazards, monitoring pesticide residues ingested through food is necessary. Few foods are consumed without processing, including washing, peeling, drying and pasteurization. During harvest, the residues left on the fruits can be carried into processed foods, such as juice, squash, jams, jelly, and raisins [37]. It is well established that food processing affects residual pesticide concentrations. Therefore, when fruits are processed, it is predicted that the residues will decay due to exposure to various processing procedures. Consequently, it is critical to include processing factors when assessing pesticide residues in processed foods.

Camara et al., [38] conducted various food processing procedures viz., cutting, washing and drying in lettuce to monitor the behavior of imidacloprid and they concluded that PF was 0.53 for imidacloprid which indicates reduction of residue content than in fresh lettuce due to food processing. Pasteurization resulted in the loss of 60.42–100% imidacloprid residues in tomato juice and paste [39, 40]. Pasteurization was found to reduce imidacloprid residues (32.45%) in strawberry juice preparation [40]. Imidacloprid residue reduction (82.66% and 66.55%) in sugared pulp and paste of winter jujube [41], 50.64 and 84.41% removal of imidacloprid residues during strawberry syrup and jam preparation, respectively [40]. Hot air over drying reduced imidacloprid residues by 70% in pomegranate, 36.73% in zucchini processing, 53% in lettuce [38, 42, 43], respectively. Processing of apples were concentrated to 0.162, 1.039, 0.102, 0.049 from 0.061, 0.372, 0.047 and 0.02 mg/kg of quinalphos, chlorpyrifos, cypermethrin and deltamethrin respectively in apple juice than in unprocessed apples [44]. Commercial processing of tomato fruits into tomato juice (under hot break) reduced 100% of imidacloprid residues [28]. Cypermethrin residues were concentrated in seedless variety of grapes to 0.46 ppm when compared to residues in fresh grapes (0.40 ppm), a study conducted by [2]. Producing apple juice from freshly harvested apples resulted in reduction of chlorpyrifos and methomyl residues by 100 and 78.1%, respectively [45].

Reddy et al., [46] studied processing effect on pesticide residues in grapes where Processing factor was calculated and was in the range of 0.01 to 0.35, 0.04 to 0.39 and 0.03 to 0.40 for juice, squash and raisin, respectively. In this study, imidacloprid was removed (59.75–67.94%) from grapes while, washing with water. Washing reduced chlorpyrifos residues (21%) in apple processing [47]. It is inferred that there is a strong correlation between water solubility (600 mg L-1) and removal of imidacloprid [48]. Crushing/homogenization does not impact residues, but it speeds up processes like hydrolysis, which releases isolated enzymes and acids from the cuticle layer more quickly, reducing residues in the juice. Partitioning characteristics of insecticide between pulp and juice are responsible for the low residual levels in juice and squash.

Clarification of juice may eliminate residues retained in the suspended particles. A negligible number of systemic insecticides might be absorbed by pulp or fruits [48]. Studies reported 93.26–97.85% removal of imidacloprid residues during processing of apples into juice [47]. Pesticide residues were significantly reduced during juice processing also reported [44, 45]. Pasteurization was found to reduce 60.42–100% of imidacloprid in tomato juice and paste [28] and imidacloprid residues (32.45%) in strawberry juice preparation [40]. During squash preparation, addition of sugar syrup to the juice reduced residues in 94.32% in present study as water was added in sugar syrup resulted in dilution of residues. Hendawi et al., [40] reported imidacloprid residue reduction during strawberry syrup (50.64%) and jam preparation (84.41%) and 66.55 and 82.66% reduction in sugared pulp and paste of winter jujube, respectively [41]. Because of evaporation and degradation, the drying process may have significantly reduced residual levels [49]. With regard to imidacloprid residues, 70.00% reduction in pomegranate by hot oven air drying [42], 53.00% in lettuce [38] and 36.73% reduction in zucchini processing [43], were reported.

In raisin preparation, residues of phosalone (68.04%) and ethion (69.55%) were removed [50]. The processing factor achieved for hexythiazox and bifenazate were 0.36 and 0.15 in grapes for raisin [51]. It is concluded that pesticides with low Kow may be removed through volatilization after drying and this is correlated with studied chemical imidacloprid where the Kow is low (0.57). Catherine et al., [52] reported that water solubility of a pesticide is an important factor during the juicing operation and pesticides with the highest water solubility were present in relatively higher amounts in the juiced carrots, tomatoes and strawberries [53]. Moreover, dimethoate is xylem mobile due to its low log Kow value of 0.7 and phloem mobile due to its pKa of 2. This is probably why washing and peeling are less effective at removing dimethoate than other organophosphates such as chlorpyrifos and parathion and thereby ending up in the filtered juice. Concentration of dimethoate quinalphos, chlorpyrifos, cypermethrin, deltamethrin were found in apple [44], chlorpyrifos in apple juice [47], in wine [54].

Pesticides residue levels were reduced during processing of food commodities but those pesticides (dimethoate, azoxystrobin, pyrimethanil) were not having a preferential partition between liquid and solid phase may be concentrated in the final processed product [55]. The poor transfer/presence of lower residues in filtered juice might be due to low water solubility (0.024 g/L) and high octanol co-efficient (Kow = 5.0) reported for emamectin benzoate, fenpropathrin and propargite in tea brewing [56, 57]. Lower residues of emamectin benzoate in grape and its processed products is might be due to high octanol co-efficient (Kow = 5.0) makes immobile in plant tissues. Our results are in line with studies where post-harvest processing and decoction of Chinese medicinal plant mugua resulted in 99.94% reduction of emamectin benzoate [58] PF of 0.06 in Chinese peony [59]. The processing of grapes into juice, squash and raisin resulted in reducing the residues as well as processing factor less than for imidacloprid, emamectin benzoate and dimethoate [46].

2. Conclusions

To reduce residue prevalence in food commodities, the governing bodies should undertake pesticide policy awareness campaigns and impose mandatory training to the grape farmers on pesticide usage, the consequences of excessive/improper pesticide use on the environment and the consumers.

Acknowledgements

The concept was perceived by Chinna Babu Naik and Malleswari Sadhineni. The book chapter was written by BKK Reddy and Maneesha. Book chapter was thoroughly reviewed by Rajareddy and Tejaswi Yelleti.

Appendices and nomenclature

PF	Processing Factor
%	Per cent
min	Minutes
P^H	Potential of Hydrogen
ppm	parts per million
g/L	grams per liter
mg L^{-1}	milligram per liter
mg/Kg	milligram per kilogram
Kow	Water Partition Coefficient
pKa	Acid dissociation Constant
O_3	Ozone

Author details

Banka Kanda Kishore Reddy[1*], Addanki Maneesha[2], Chinna Babu Naik[3], Malleswari Sadhineni[1], Tejaswi Yelleti[4] and G. Raja Reddy[5]

1 Krishi Vigyan Kendra, Acharya NG Ranga Agricultural University, Guntur, India

2 Department of Entomology, Sri Krishnadevaraya College of Agricultural Sciences, Ananthapuramu, Andhra Pradesh, India

3 Indian Institute of Rice Research, Hyderabad, India

4 National Institute of Plant Health Management, Hyderabad, India

5 Indian Agricultural Research Institute, New Delhi, India

*Address all correspondence to: bankakishorereddy@gmail.com

References

[1] Xie X, Gong S, Wang X, Wu Y, Zhao L. Simplified RP-HPLC method for multi-residue analysis of abamectin, emamectin benzoate and ivermectin in rice. Food Additives and Contaminants: Part A. 2011;**28**(1):19-25

[2] Lentza-Rizos C, Kokkinaki K. Residues of cypermethrin in field-treated grapes and raisins produced after various treatments. Food Additives & Contaminants. 2002;**19**(12):1162-1168. DOI: 10.1080/026520302100001238 5

[3] Christensen HB, Granby K, Rabolle M. Processing factors and variability of pyrimethanil, fenhexamid and tolylfluanid in strawberries. Food Additives and Contaminants. 2003;**20**:728-741

[4] Elgueta S, Valenzuela M, Fuentes M, Meza P, Manzur JP, Liu S, et al. Pesticide residues and health risk assessment in tomatoes and lettuces from farms of metropolitan region Chile. Molecules. 2020;**25**(2):355

[5] Bempelou E, Anagnostopoulos C, Kiousi M, Malatou P, Liapis K, Kouloussis N, et al. Temporal variation in pesticide residues in citrus fruits from Chios, Greece, before and after the development of an integrated Pest management strategy (IPMS): A five-year study (LIFE13 ENV GR/000414). Toxics. 2021;**9**(12):323

[6] Abd-Elhaleem ZA. Pesticide residues in tomato and tomato products marketed in Majmaah province, KSA, and their impact on human health. Environmental Science and Pollution Research International. 2020;**27**(8):8526-8534

[7] Pirsaheb M, Fattahi N, Karami M, Ghaffari HR. Simultaneous determination of deltamethrin, permethrin and malathion in stored wheat samples using continuous sample drop flow microextraction followed by HPLC–UV. Journal of Food Measurement and Characterization. 2018;**12**(1):118-127

[8] Shokrzadeh M, Karami M, Ghadi MA. Measuring organophosphorus insecticide residue in rice produced in Amol, north of Iran. Journal of Mazandaran University of Medical Sciences. 2013;**23**(1):215-221

[9] Pirsaheb M, Fattahi N, Rahimi R, Sharafi K, Ghaffari HR. Evaluation of abamectin, diazinon and chlorpyrifos pesticide residues in apple product of Mahabad region gardens: Iran in 2014. Food Chemistry. 2017;**231**:148-155

[10] Jalalizand AR, Fakhari H, Modaresi M, Shayeghi M, Abtahi M. The amount of dursban pesticide residues in Isfahan sugar beet. Procedia Environmental Sciences. 2011;**8**:235-239

[11] Jayakrishnan S, Dikshit AK, Singh JP, Pachauri DC. Dissipation of lambda-cyhalothrin on tomato (Lycopersicon esculentum mill.) and removal of its residues by different washing processes and steaming. Bulletin of environmental contamination and toxicology. 2005;**75**(2):324-328

[12] Akyildiz A. Effects of various rinsing conditions on residues of chlorpyrifos ethyl, acetamiprid and penconazole in grapes. Journal of Agricultural Sciences. 2014;**20**(2):112-119

[13] Harinathareddy A, Prasad NB, Devi KL. Effect of household processing methods on the removal of pesticide residues in tomato vegetable. Journal of Environmental Research and Development. 2014;**9**(1):50

[14] Ganguli A, Rao P. A study on decontamination of pesticides in grapes. International Journal of Applications and Biology Pharm. 2017;**8**:31-37

[15] Awasthi MD. Decontamination of insecticide residues on mango by washing and peeling. Journal of Food science and Technology. 1993;**30**(2):132-133

[16] Devi M, Duhan A, Kumari B, Yadav GS. Determination of dimethoate, lambda-cyhalothrin and malathion residues in guava fruits using GCMS-tandem mass spectrometry. Indian Journal of Horticulture. 2016;**73**(2):197-201. DOI: 10.5958/0974-0112.2016.00047.5

[17] Cabras P, Angioni A. Pesticide residues in grapes, wine, and their processing products. Journal of Agricultural and Food Chemistry. 2000;**48**:967-973. DOI: 10.1021/jf990727a

[18] Reddy DJ, Rao BN. Dissipation and decontamination of chlorpyrifos and quinalphos residues on/in grape berries. Pest Management in Horticultural Ecosystems. 2002;**8**(2):103-108

[19] Reddy DJ, Rao BN. Dissipation and decontamination of carbaryl and isoprocarb residues on grapes. Indian Journal of Entomology. 2003;**65**(1):62-66

[20] Reddy DJ, Rao BN. Dissipation and decontamination of bifenthrin residues on grapes. Pestology. 2005;**25**(8):33-35

[21] Aguilera A, Valverde A, Camacho F, Boulaid M, Garcia-Fuentes L. Household processing factors of acrinathrin, fipronil, kresoxim-methyl and pyridaben residues in green beans. Food Control. 2014;**35**:146-152. DOI: 10.1016/j.foodcont.2013.06.038

[22] Srinivasa RS, Narendra RC, Shashi V. Decontamination methods utilising house hold practices for removing pesticides on field bean for food safety. Journal of Nutrition and Health Food Engineering. 2018;**8**(3):260-267

[23] Kaushik E, Dubey JK, Patyal SK, Katna S, Chauhan A, Devi N. Persistence of tetraniliprole and reduction in its residues by various culinary practices in tomato in India. Environmental Science and Pollution Research. 2019;**26**(22):22464-22471

[24] Srivastava A, Chabra A, Singh GP, Srivastava PC. Efficacy of different decontamination processes in mitigation of pesticide residues from chili crop. Journal of Food Protection. 2021;**84**(5):767-771

[25] Vijayasree V, Bai H, Mathew TB, George T, Xavier G, Kumar NP, et al. Dissipation kinetics and effect of different decontamination techniques on the residues of emamectin benzoate and spinosad in cowpea pods. Environmental Monitoring and Assessment. 2014;**186**(7):4499-4506. DOI: 10.1007/s10661-014-3714-9

[26] Acoglu B, Omeroglu PY. Effectiveness of different type of washing agents on reduction of pesticide residues in orange (Citrus sinensis). LWT-Food Science and Technology. 2021;**147**:111690

[27] Reddy AH, Prasad NBL, Devi KL, Raveendranath D, Ramesh B. Risk mitigation methods on removal of pesticide residues in grape fruits for food safety. Research Journal of Pharmaceutical Biological and Chemical Sciences. 2015;**6**(2):1568-1572

[28] Romeh AA, Mekky TM, Ramadan RA, Hendawi. Dissipation of Profenofos, Imidacloprid and Penconazole in tomato fruits and products. Bulletin of Environmental Contamination and Toxicology.

2009;**83**:812-817. DOI: 10.1007/s00128-009-9852-z

[29] Nowowi MFM, Ishak MAM, Ismail K, Zakaria SR. Study on the effectiveness of five cleaning solutions in removing chlorpyrifos residues in cauliflower (Brassica oleracea). Journal of Environmental Chemistry and Ecotoxicology. 2016;**8**(7):69-72

[30] Pandiselvam R, Kaavya R, Jayanath Y, Veenuttranon K, Piraya Lueprasitsakul V, Divya AK, et al. Ozone as a novel emerging technology for the dissipation of pesticide residues in foods–a review. Trends in Food Science & Technology. 2020;**97**:38-54, ISSN 0924-2244,. DOI: 10.1016/j.tifs.2019.12.017

[31] Ong KC, Cash JN, Zabik MJ, Siddiq M, Jones AL. Chlorine and ozone washes for pesticide removal from apple and processed apple sauce. Food Chemistry. 2009;**55**(2):153-160

[32] Ikeura H, Kobayashi F, Tamaki M. Removal of residual pesticides in vegetables using ozone microbubbles. Journal of Hazardous Materials. 2011b;**186**:956-959

[33] Whangchai K, Uthaibutra J, Phiyanalinmat S, Pengphol S, Nomura N. Effect of ozone treatment on the reduction of chlorpyrifos residues in fresh lychee fruits. Ozone. Science and Engineering. 2011;**33**:232-235

[34] Kusvuran E, Yildirim D, Mavruk F, Ceyhan M. Removal of chloropyrifos ethyl, tetradifon and chlorothalonil pesticide residues from citrus by using ozone. Journal of Hazardous Materials. 2012;**242**:287-300

[35] Lozowicka B, Jankowska M, Hrynko I, Kaczynski P. Removal of 16 pesticide residues from strawberries by washing with tap and ozone water, ultrasonic cleaning and boiling. Environmental Monitoring and Assessment. 2016;**188**:51-55

[36] Velioglu YS, Cönger E, Aksu P, Fİkİrdeşici-Ergen S, Yİğİt-Baltacı HM. Effects of ozonated water washing on pesticide removal, ascorbic acid and colour of tomatoes. GIDA-Journal of Food. 2016;**41**(5):337-344

[37] Shabeer ATP, Girame R, Hingmire S, Banerjee K, Sharma AK, Oulkar D, et al. Dissipation pattern, safety evaluation, and generation of processing factor (PF) for pyraclostrobin and metiram residues in grapes during raisin preparation. Environmental Monitoring and Assessment. 2015;**187**(2):1-8. DOI: 10.1007/s10661-015-4268-1

[38] Camara MA, Barba A, Cermeño S, Martinez G, Oliva J. Effect of processing on the disappearance of pesticide residues in fresh-cut lettuce: Bioavailability and dietary risk. Journal of Environmental Science and Health, Part B. 2017;**52**(12):880-886. DOI: 10.1080/03601234.2017.1361767

[39] Lopez-Fernandez O, Rial-Otero R, Simal-Gandara J. Factors governing the removal of mancozeb residues from lettuces with washing solutions. Food Control. 2013;**34**:530-538. DOI: 10.1016/j.foodcont.2013.05.022

[40] Hendawi MY, Romeh AA, Mekky TM. Effect of food processing on residue of imidacloprid in strawberry fruits. Journal of Agriculture, Science and Technology. 2010;**15**:951-959. DOI: 10.21608/JPD.2010.42411

[41] Peng W, Zhao L, Liu F, Xue J, Li H, Shi K. Effect of paste processing on residue levels of imidacloprid, pyraclostrobin, azoxystrobin and fipronil in winter jujube. Food

Additives and Contaminants, Part A. 2014;**31**:1562-1567. DOI: 10.1080/19440049.2014.941948

[42] Utture SC, Banerjee K, Kolekar SS, Dasgupta S, Oulkar DP, Patil SH, et al. Food safety evaluation of buprofezin, dimethoate and imidacloprid residues in pomegranate. Food Chemistry. 2012;**131**(3):787-795. DOI: 10.1016/j.foodchem.2011.09.044

[43] Oliva J, Cermeno S, Camara MA, Martinez G, Barba A. Disappearance of six pesticides in fresh and processed zucchini, bioavailability and health risk assessment. Food Chemistry. 2017;**229**:172-177. DOI: 10.1016/j.foodchem.2017.02.076

[44] Rasmusssen RR, Poulsen ME, Hansen HCB. Distribution of multiple pesticide residues in apple segments after home processing. Food Additives and Contaminants. 2003;**20**(11):1044-1063. DOI: 10.1080/02652030310001615221

[45] Zabik MJ, El-Hadidi MFA, Cash JN, Zabik ME, Jones AL. Reduction of azinphos-methyl, chlorpyrifos, esfenvalerate, and methomyl residues in processed apples. Journal of Agricultural and Food Chemistry. 2000;**48**(9):4199-4203. DOI: 10.1021/jf9913559

[46] Reddy BKK, Bhuvaneswari K, Geetha P, Thamilarasi N, Suganthi A, Paramasivam M. Effect of decontamination and processing on insecticide residues in grape (Muscat Hamburg). Environmental Science and Pollution Research. 2022;**29**(50):75790-75804

[47] Kong Z, Shan W, Dong F, Liu X, Xu J, Li M, et al. Effect of home processing on the distribution and reduction of pesticide residues in apples. Food Additives & Contaminants. 2012;**Part A, 29**(8):1280-1287. DOI: 10.1080/19440049.2012.690347

[48] Malhat F, Bakery M, Anagnostopoulos C, Youssef M, Abd El-Ghany W, Abdallah A, et al. Investigation of the dissipation behaviour and exposure of spitotetramat, flonicamid, imidacloprid and pymetrozine in open field strawberries in Egypt. Food Additives & Contaminants. 2021;**Part A, 38**(12):2128-2136

[49] Bajwa U, Sandhu KS. Effect of handling and processing on pesticide residues in food-a review. Journal of food science and technology. 2014;**51**(2):201-220

[50] Rahimi A, Heshmati A, Nili-Ahmadabadi A. Changes in pesticide residues in field-treated fresh grapes during raisin production by different methods of drying. Drying Technology. 2021;**40**(8):1-14. DOI: 10.1080/07373937.2021.1919140

[51] Shabeer ATP, Girame R, Hingmire S, Jadhav M, Jain P. Residue dissipation, evaluation of processing factor and safety assessment of hexythiazox and bifenazate residues during drying of grape to raisin. Environmental Science and Pollution Research. 2020;**27**(33):41816-41823

[52] Burchat CS, Ripley BD, Leishman PD, Ritcey GM, Kakuda Y, Stephenson GR. The distribution of nine pesticides between the juice and pulp of carrots and tomatoes after home processing. Food Additives & Contaminants. 1998;**15**(1):61-71. DOI: 10.1080/02652039809374599

[53] Han Y, Li W, Dong F, Xu J, Liu X, Lie Y. The behavior of chlorpyrifos and its metabolite 3,5,6-trichloro-2- pyridinol in tomatoes during home canning. Food

Control. 2013;**31**:560-565. DOI: 10.1016/j.foodcont.2012.11.050

[54] Pazzirota T, Martin L, Mezcua M, Ferrer C, Fernandez-Alba AR. Processing factor for a selected group of pesticides in a wine-making process: Distribution of pesticides during grape processing. Food Additives & Contaminants. 2013;**Part A, 30**(10):1752-1760. DOI: 10.1080/19440049.2013.815806

[55] Cus F, Cesnik HB, Bolta SP, Gregorcic A. Pesticide residues in grapes and vinification process. Food Control. 2010;**21**:1512-1518

[56] Manikandan N, Seenivasan S, Ganapathy MNK, Muraleedharan NN, Selvasundaram R. Leaching of residues of certain pesticides from black tea to brew. Food Chemistry. 2009;**113**(2):522-525

[57] Zhou L, Luo F, Zhang X, Jiang Y, Lou Z, Chen Z. Dissipation, transfer and safety evaluation of emamectin benzoate in tea. Food chemistry. 2016;**202**:199-204

[58] Xiao JJ, Wang F, Ma JJ, Xu X, Liao M, Fang QK, et al. Acceptable risk of fenpropathrin and emamectin benzoate in the minor crop Mugua (Chaenomeles speciosa) after postharvest processing. Environmental Pollution. 2021a;**276**:116716

[59] Xiao J, Ma J, Wang F, Xu X, Liao M, Shi Y, et al. Effect of decocting on the pesticide residues in Paeoniae radix lactiflora and corresponding exposure risk assessment. Environmental Science and Pollution Research. 2021b;**28**(13):16655-16662

Chapter 8

Determination of Insecticidal and Larvicidal Activities of *Artemisia herba-alba* Essential Oil against *Tribolium confusum*

Bouzenna Hafsia, Dorsaf Ben Jeddou, Hadil Kadri, Guesmi Fatma, Abassi Raoudha and Najla Hfaiedh

Abstract

The present work was to evaluated the insecticidal activity of the essential oil extracted from an aromatic plant *Artemisia herba alba* (Lamiaceae) on larvaes and insects of *Tribolium confusum*. The results obtained showed that the essential oil *A. herba alba* revealed significant antiradical activities against the DPPH radical and a powerful reducing power. The repellent effect and the toxicity by contact were tested at different concentrations (2, 4, 6, 8 μl/ml). The results revealed that the essential oil of *A.herba alba* was moderately toxic after 24 h towards the insects and the larvaes. While the repellent effect showed that *Artemisia oil* had a very strong repellent effect on the larvae's. These results added to those obtained by other authors on the usefulness and effectiveness of essential oils in the control of different.

Keywords: *Artemisia herba alba*, essential oils, toxicity, *Tribolium confusum*, repellent effect

1. Introduction

Plants represent an immense source of complex chemical molecules exploited by man in the perfume, food, cosmetics and pharmaceutical industries. Most plants contain essential oils; they are then called "aromatic plants". These essential oils were found in many parts of the plant: wood, leaves, fruits, bark, seeds and roots. These are complex mixtures made up of several dozen or even more than a hundred compounds, mainly terpenes and aromatic compounds.

In addition, plant production of economic interest often suffered damages caused by pests, disease agents such as viruses, bacteria, insects or fungi. To control the crop enemies, producers had often used synthetic chemicals, including insecticides, fungicides, bactericides and herbicides. Unfortunately, the application of these chemicals had caused unexpected problems both in terms of soil contamination by non-degradable chemical molecules and on human and animal health by the harmful residues that persist on food.

In order to limit these drawbacks, several academic and public laboratories had taken an interest in the development of biological control [1]. In this work, was interested in a medicinal plant, it is *Artemisia herba-alba*, for which, it's essential oils was extracted, and evaluated its possible interest in the context of biological control.

A. herba-alba has been used in traditional medicine by many cultures since ancient times. It is a very popular remedy that is often used to facilitate digestion, calm abdominal pain and certain liver ailments. Its roots were indicated against certain nervous disorders [2]. Their aqueous extract were traditionally used in Jordan as an antidote against the venoms of several types of snakes and scorpions. In North Africa, it was known to treat bronchitis, abscess, diarrhea and as a vermifuge. Its oil essential had many activities including antioxidant, hypoglycemic, antifungal, antibacterial [3, 4].

In this context, the study was evaluated the bioinsecticide effect of the essential oil of *A. herba-alba* on adults and larvae of *T. confusum*.

2. Materials and methods

2.1 Extraction of essential oil from *A. herba alba*

2.1.1 Plant material

The biomass used for the extraction of essential oils was composed of the aerial parts of *A. herba alba*. They were collected from Jbel Mdhila mountain of the Gafsa region in April 2022. The aerial parts of this plant were dried in the open air, away from light, for 8 days. After drying, the dried plant material was ground in a mortar to increase the solvent-sample contact surface.

2.1.2 Chemicals

Acetone, DPPH, BHT, potassium ferricyanide, trichloroacetic acid were purchased from Sigma-Aldrich Chemicals Co. (St. Louis, MO, USA).

2.2 Extraction of volatile oils

The extraction of essential oils was done by the method of hydrodistillation: Hydro-distillation is a technique for extracting essential oils that is done using a simple distillation. In a 1 liter flask, 500 mL of distilled water was added with a few pumice stones to regulate the heating and about 50 g of the dried leaves of plant. The water was brought to a boil, taking care not to heat until completely dry. Thus, the water vapor entrains the volatile organic substances which condense in the refrigerator then flow drop by drop and are collected at the other end of the assembly in the graduated cylinder. The volume of the distillate is approximately 130 mL.

2.3 Determination of extraction yield

The yield of essential oils is defined as being the ratio between the mass of the essential oil obtained and the mass of the plant material treated.

$$R = \left(\frac{m}{m0}\right) \times 100 \quad (1)$$

With R: essential oil yield (%), m: mass in grams of the essential oil, m0: mass in grams of dry plant matter.

2.4 Study of biological activities

2.4.1 Antioxidant activity of essential oil.

The antioxidant activity was tested using two methods: scavenging of the free radical DPPH and determination of its reducing power (FRAP).

2.4.2 The DPPH• (2,2-diphenyl-1-picrylhydrazyl) test

DPPH• (2,2-Diphenyl-1-picrylhydrazyl) is a stable purplish-colored free radical that absorbs at 517 nm. In the presence of anti-radical compounds, the DPPH• radical is reduced and changes color by turning yellow. This method is based on measuring the capacity of antioxidants to scavenge the 2,2-diphenyl-1-picrylhydrazil (DPPH•) radical. The latter is reduced to the form of hydrazine (non-radical) by accepting a hydrogen atom. The effect of oils on DPPH is measured by the procedure described by Öztürk et al. [5]. The percentage of DPPH anti-radical activity is calculated according to the equation:

2.4.3 Reducing power (FRAP)

The reducing power of ferric ions ($Fe3+$) of an extract was evaluated using the method described by Oyaizu [6]. A volume of 1.25 mL of a phosphate buffer solution (0.1 M, pH 6.6) and 1.25 mL of potassium ferricyanide (1% w/v) were mixed with 500 µL of an extract solution at concentrations variables. The reaction mixture was incubated for 30 minutes at 50°C, then a volume of 1.25 mL of 10% w/v trichloroacetic acid was added to stop the reaction, the mixture was centrifuged at 1500 × g for 10 min. A volume of 1.25 mL of the supernatant is mixed with 1.25 mL of distilled water and 250 µL of a 0.1% w/v solution of FeCl3 then incubated for 10 minutes. The absorbance of the mixture was measured at 700 nm. The greater the absorbance of the reaction mixture, the greater the reducing power. BHT was used as a positive control with the same concentrations (50–300 µg/mL) and under the same experimental conditions was used as a standard antioxidant and the results are presented as the change in absorbance with concentration. From the graph the effective concentration (EC50) is determined which gives an absorbance of 0.5. The lower the EC50, the stronger the reducing power. The test measuring the reducing power was repeated in three trials for each extract.

2.5 Study of the insecticidal and larvicidal activity of the essential oil of *A. herba alba*

The toxicity tests of *A. herba-alba* on insects and larvae of *T. confusum* were carried out according to two modes of penetration, one penetration by contact and the other by repellent effect.

2.6 Evaluation of the toxicity of this oil by contact effect

After preparing the doses, each solution was spread evenly on a disc of Wattman-type filter paper previously placed in Petri dishes of the same diameter. After evaporation of the dilution solvent, all the boxes are infested by 10 insects or larvae of *T. confusum*. We carried out three repetitions for the four doses of essential oil tested, and the same for the control not treated with the essential oil. A count of the dead insects is carried out after 24 hours of this treatment.

2.7 Test to evaluate the repellent effect of essential oil on insects and larvaes

The repellent effect of essential oils against adults and larvae of *T. confusum* was evaluated using the technique described by McDonald et al. [7] using the preferential area method on filter paper. In this experiment, the 9 cm diameter filter paper discs used for this purpose were divided into two equal parts. Four different essential oil contents were used in this test: 2, 4, 6, 8 µl/ml [8]. 0.5 mL of each dose was placed on one half of the disc, while the other half received only acetone. The two discs were placed in the open air for 10 minutes, in order to evaporate the acetone. After acetone drying. The filter paper disc was placed in a 9 cm diameter Petri dish. A batch of 10 individuals of *T. confusum* were placed in the center of filter paper. The boxes are placed under ambient temperature conditions to calculate the percentage of repellent value for each essential oil. The percentage of repulsion is thus calculated according to the following formula:

$$\text{Percent Repulsion}\,(PR\%) = ((NC - NT)/\text{NC}) \times 100 \qquad (2)$$

Where: NC: number of the individual present on the part of the disc treated only with acetone NT: number of the individual present on the part of the disc treated with the prepared dose. The average percentage repellency for each oil is calculated (PR) and assigned to one of the different repellent classes varying from 0 to V [7] which showed in **Table 1**.

2.8 Data analysis method

2.8.1 Mortality correction by Abbott's method

The effectiveness of this essential oil is evaluated by mortality. The results of the tests carried out do not only represent the mortality caused by the oil but there is also the natural mortality. This mortality is corrected using the Abbott formula.

$$MC(\%) = \left((MT - Mt)/(100 - Mt)\right) \times 100 \qquad (3)$$

_ MC %: percentage of corrected mortality.
_ Mt.: percentage of mortality obtained in the control population.
_ MT: percentage of mortality obtained in the treated population.

Class	Repulsion interval	Property of the treated substance
Class 0	PR ≤ 0,1%	Not repellent
Class I	0,1 < PR ≤ 20%	Very weakly
Class II	20 < PR ≤ 40%	Repellent weakly
Class III	40 < PR ≤ 60%	Repellent moderately
Class IV	60 < PR ≤ 80%	Repellent
Classe V	80 < PR ≤ 100%	Very repellent

Table 1.
Percentage of repulsion (PR) according to the classification of Donald et al. [7].

3. Results and discussion

3.1 Aspect and yield of the essential oil

The extraction by hydrodistillation of the aerial parts *of A. herba-alba*, obtained the essential oil of yellow color and very strong smell. The yield of the essential oil obtained in relation to the total weight of the dry plant is 0.16%. The yield was low compared to that obtained by the same plant in different regions of Algeria (from 0.2 to 0.95%) [2, 9]. That of Spain was 0.41 to 2.30% [10] and that of Jordan 1.3% [11]. This difference in yields can be attributed not only to the geographical origin of the plant, but also to many factors such as: growth stage, pedoclimatic conditions, place of production, etc. Several studies had shown the influence of the vegetative cycle and the extraction technique on the yield and quality of EO.

3.2 Evaluation of antioxidant activity

In order to evaluate the antioxidant activity of *A. herba alba* essential oil, different antioxidant tests (DPPH and FRAP) were used.

3.2.1 Anti-radical activity against DPPH radical

The antioxidant effect was assessed by the DPPH test as the ability of DPPH •, compounds with antiradical activity, to scavenge free radicals. From these results, it deduced that the evolution of the anti-free radical activity is studied in vitro against the radical DPPH is dose-dependent (**Figure 1**). Indeed, the increase in the concentration of the oil is followed by an increase in the anti-DPPH activity. These results were also expressed by IC50. Indeed, the IC50 parameter is defined as being the concentration of the substrate which leads to a loss of 50% of the activity. Indeed, the oil with the lowest IC50 value exerts the most powerful antiradical activity. Indeed, the oil reported a value of IC50 = 160 µg/mL. This reflected a powerful antioxidant power by comparing it with the standard (ascorbic acid) which has the IC50 value = 210 µg/mL.

3.2.2 Iron reduction test (FRAP)

The reducing power was based on the reduction of the ferric ion ($Fe3+$) to ferrous ion ($Fe2+$) by transfer of an electron or donation of a hydrogen atom.

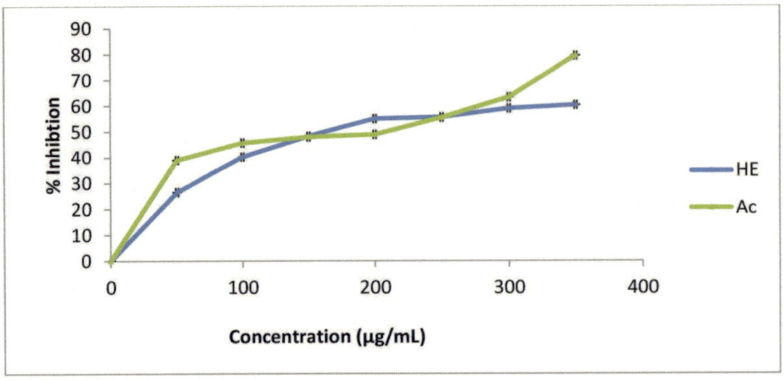

Figure 1.
Antiradical activity of the essential oil of Artemisia herba alba. Each value represents the mean ± SD (n = 3 trials for each sample). HE: Essential oil, Ac: ascorbic acid.

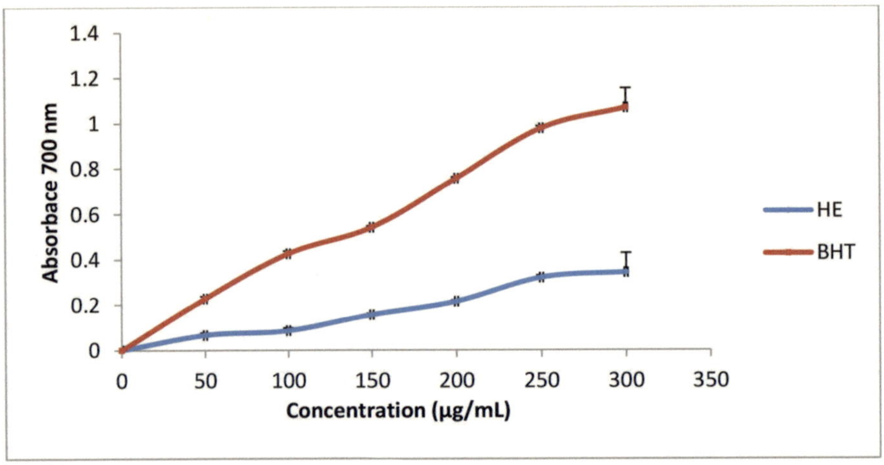

Figure 2.
Graphic representation of the reducing power of A. herba alba *and BHT each value represents the mean ± SD (n = 3 trials for each sample).*

The **Figure 2** represented the reducing power of essential oil of *A. herba alba* and the BHT standard. The results showed that the reducing power increased with the concentration. The absorbance at 700 nm of the oil increased proportionally to its concentration. However, in comparison with BHT (used as a control), the reducing power of this oil was lower. Indeed, even at high concentration (300 µg/mL), we observed an absorbance of 0.32 for oil against 1.00 for BHT.

3.2.3 Insecticidal and larvicidal activity of the essential oil of A. herba alba

To evaluate the insecticidal and larvicidal effect of the essential oil of *A. herba alba* the mortality rate of adults and larvae's by contact and repellent method were estimated.

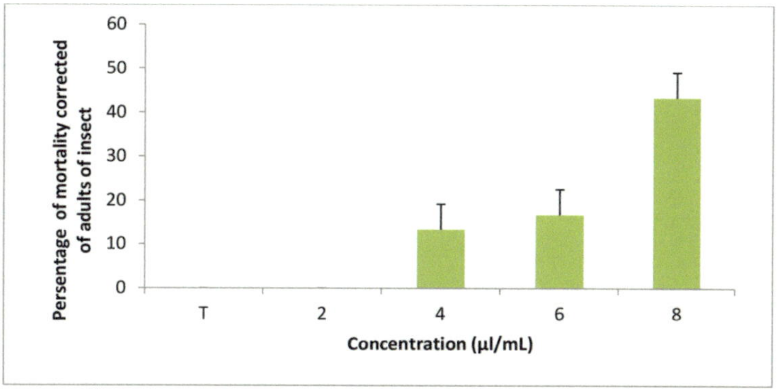

Figure 3.
Corrected mortality percentage of T. confusum insects after 24 h of exposure to A.herba alba *essential oil. Each value represents the mean ± SD (n = 3 replicates with 10 insects each).*

3.2.3.1 Assessment of adult and larval mortality by contact effect

The corrected mortalities were mentioned in **Figures 3** and **4**. The results indicated that the larvae were more resistant than the insects. The results showed an increase in the concentration-corrected mortality rate. As an aromatic plant, the volatility of the essential oil due to the montepenes and the major Camphor compound of this oil gives it an effectiveness in controlling stock insect pests. These results confirm other inferences reporting Chaied et al. [12] that this oil showed good insecticidal activity.

3.2.3.2 Determination of the repellent activity of the essential oil

The results of the evaluation of the repellent effects of essential oil of *A. herba alba* on the larvae and adults of *T. confusum* were presented in **Tables 2** and **3**. The percentage of repellent essential oil used increases according to the dose and time.

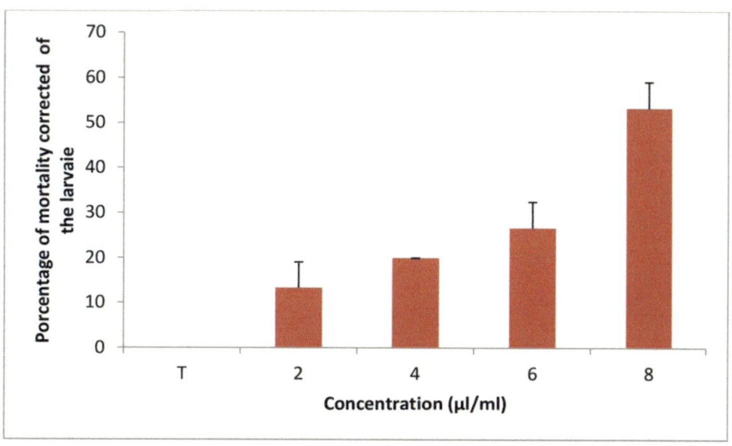

Figure 4.
Percentage of corrected mortality of larvae of T. confusum *after 24 h of exposure to* A.herba alba *essential oil. Each value represents the mean ± SD (n == 3 replicates with 10 insects each).*

	30 min		60 min		90 min		120 min	
	Repellent effects	**Classe**	**Repellent effects**	**Classe**	**Repellent effects**	**Classe**	**Repellent effects**	**Classe**
2	13.3 ± 8.08	Class I	20 ± 6	Class I	46.66 ± 3.05	Class III	53.33 ± 3.05	Class II
4	20 ± 4	Class I	36.03 ± 3.05	Class II	42.2 ± 3.4	Class III	60 ± 4	Class III
6	46.66 ± 4.6	Class III	62.5 ± 3.03	Class IV	73.33 ± 4.6	Class IV	100 ± 00	Classe V
8	53.33 ± 1.1	Class III	64.4 ± 3.35	Class IV	73.33 ± 4.6	Class IV	100 ± 00	Classe V

Table 2.
Repellent effects on T. confusum *larvae exposed to* A. herba alba *essential oil.*

	30 min		60 min		90 min		120 min	
	Effet répulsif	**Classe**	**Effet répulsif**	**Classe**	**Effet répulsif**	**Classe**	**Effet répulsif**	**Classe**
2	6	Class I	80 ± 1.15	Class IV	80 ± 3.4	Class IV	77.77 ± 3.84	Class IV
4	80 ± 1.15	Classe V	80 ± 1.15	Classe V	80 ± 2	Classe V	100 ± 00	Classe V
6	80 ± 2	Classe V	93.33 ± 1.15	Classe V	93.33 ± 1.15	Classe V	100 ± 00	Classe V
8	86.66 ± 2.3	Classe V	100 ± 00	Classe V	100 ± 00	Classe V	100 ± 00	Classe V

Table 3.
Repellent effects on adults of Tribolium confusum *exposed to the essential oil of* A. herba alba.

Our deductions revealed that the highest dose of *A. herba alba* had a strong repellent effect of 100% on *T. confusum* larvaes and adults respectively. From these results, it was concluded that higher concentration of essential oil resulted in maximum pest repellency compared to lower concentrations. According to the classification established by Mc. Donald et al. [7], it can be concluded that the oil recorded a very strong repellent effect against adults and larvae of *T. confusum*. Many researchers have pointed out that certain plant essential oils exhibited strong repellent effects against harmful storage [13–15].

4. Conclusion

The use of bio-insecticides was a major and effective potential for crop protection. The use of essential oils extracted from aromatic plants has demonstrated their insecticidal activity against insect pests of stocks. In this work, initially, the evaluation of the antioxidant activity by the DPPH free radical scavenging method and the FRAP test showed that the essential oil has a significant antioxidant power. On the other hand, treatment with *A. herba alba* essential oil on *T. confusum* larvae and insects gave a repellent effect and a toxic effect. At the end of this study, the results obtained are in agreement with previous results. They largely confirm the usefulness of essential oils in the pest control program, particularly *in P. interpunctella* and *E. kuheniella* and in different orders of insects.

Funding

This research was supported and funded by the Ministry of Higher Education and Scientific Research (MHESR, Tunisia).

Conflict of interest

The authors declare that they have no conflict of interest.

Author details

Bouzenna Hafsia*, Dorsaf Ben Jeddou, Hadil Kadri, Guesmi Fatma, Abassi Raoudha and Najla Hfaiedh
Laboratory of Biotechnology and Biomonitoring of the Environment and Oasis Ecosystems (LBBEEO), Faculty of Sciences of Gafsa, University of Gafsa, Tunisia

*Address all correspondence to: bouzenna.hafsia@hotmail.fr

References

[1] Amri I, Hamrouni L, Mohsen H, Jamoussi B, Lebdi K. Essential oils as biological alternatives to protect datepalm (*Phoenix dactylifera* L.) against Ectomyeloisceratoniae. Journal of Agricultural Research. 2014;**74**:273-279. DOI: 10.4067/S0718-58392014000300004

[2] Bouzidi N, Mederbal K, Raho BG. Antioxidant activity of essential oil of *Artemisia herba alba*. Journal of Applied Environmental and Biological Sciences. 2016;**6**(5):59-65

[3] Taştekin D, Atasever M, Adigüzel G, Keleş M, Taştekin A. Hypoglycaemic effect of artemisia herba-alba in experimental hyperglycaemic rats. Bulletin- Veterinary Institute in Pulawy. 2006;**50**:235-238

[4] Heleili N, Merradi M, Oucheriah Y, Belkadi S, Ayachi A, Adjroud I. Antimicrobial activity of essential oil of Artemisia herbaalba Asso from Eastern Algeria. 2018;**12**:(5)361-366

[5] Öztürk M, Aydoğmuş-Öztürk F, Duru ME, Topçu G. Antioxidant activity of stem and root extracts of rhubarb (Rheum ribes): An edible medicinal plant. Food Chemistry. 2007;**103**(2):623-630. DOI: 10.1016/j.foodchem.2006.09.005

[6] Oyaizu M. Studies on products of browning reaction. The Japanese Journal of Nutrition and Dietetics. 1986;**44**(6):307-315

[7] McDonald LL, Guy RH, Speirs RD. Prelminary evaluation of new candidate materials as toxicants, repellents and attractants against stored product insects. Marketing Research Raport of Agriculture. 1970;**8**:398-494. DOI: 10.22004/ag.econ.312345

[8] Arab R, Lemeailbi N, Benhissen S. Repellent activity of essential oils from *Artemisia Herba Alba* Asso and *Teucrium Polium L.* against tow stored product insects. NVEO-natural Volatiles & Essential Oils Journal. 2022;**9**(10):1508-1515

[9] Ghanmi M, Satrani B, Aafi A, Isamili MR, Houti H, El Monfalouti H, et al. Effet de la date de récolte sur le rendement, la composition chimique et la bioactivité des huiles essentielles de l'armoise blanche (*Artemisia herba-alba*) de la région de Guerçif (Maroc oriental). Phytothérapie. 2010;**8**:295-301. DOI: 10.1007/s10298-010-0578-1

[10] Salido S, Altarejos J, Nogueras M, Sánchez A. Chemical composition of the essential oil of Artemisia herba-alba Asso ssp valentina (Lam.) Marcl. Journal of Essential Oil Research. 2001;**13**(4):221-224. DOI: 10.1080/10412905.2006.9699206

[11] Hudaiba M, Aburjai T. Composition of the essential oil from Artemisia herba-alba grown in Jordan. Journal of Essential Oil Research. 2006;**18**(3):301-304. DOI: 10.1080/10412905.2006.9699096

[12] Chaieb I, Ben Hamouda A, Tayeb W, Zarrad K, Bouslema T, Laarif A. The Tunisian artemisia essential oil for reducing contamination of stored cereals by *Tribolium castaneum*. Food Technology and Biotechnology. 2018;**56**(2):247-256. DOI: 10.17113/ftb.56.02.18.5414

[13] Martínez LC, Plata-Rueda A, Colares HC, Campos JM, Dos Santos MH, Fernandes FL, et al. Toxic effects of two essential oils and theirconstituents on the meal worm beetle, Tenebriomolitor. Bulletin of Entomological Research.

2018;**108**:716-725. DOI: 10.1017/S0007485317001262

[14] Mishra BB, Tripathi SP, Tripathi CPM. Repellent effect of leaves essential oils from Eucalyptus globulus (Mirtaceae) and Ocimum basilicum (Lamiaceae) against two major stored grain insect pests of coleopterons. Journal of Nature and Science. 2012;**10**(2):50-54 Available from: http://www.sciencepub.net/nature

[15] Moutassem D, Bellik Y, Sennef MEH. Toxicity and repellent activities of *Thymus pallescens* and *Cymbopogoncitratus* essential oils against *Sitophylus granarius*. Plant Protection Science. 2021;**57**:297-309. DOI: 10.17221/185/2020-PPS